Lecture Notes in Mathematics

Edited by A. Dold and B. Eckmann

663

J. F. Berglund
H. D. Junghenn
P. Milnes

Compact Right Topological Semigroups and Generalizations of Almost Periodicity

Springer-Verlag
Berlin Heidelberg New York 1978

Authors
John F. Berglund
Virginia Commonwealth University
Richmond, Virginia 23284/USA

Hugo D. Junghenn
George Washington University
Washington, D.C. 20052/USA

Paul Milnes
The University of Western Ontario
London, Ontario
Canada N6A 5B9

AMS Subject Classifications (1970): 22A15, 22A20, 43A07, 43A60

ISBN 3-540-08919-5 Springer-Verlag Berlin Heidelberg New York
ISBN 0-387-08919-5 Springer-Verlag New York Heidelberg Berlin

2141/3140-543210

INTRODUCTION

The primary objective of this monograph is to present a reasonably self-contained treatment of the theory of compact right topological semigroups and, in particular, of semigroup compactifications. By semigroup compactification we mean a compact right topological semigroup which contains a dense continuous homomorphic image of a given semitopological semigroup. The classical example is the Bohr (or almost periodic) compactification (a,AR) of the usual additive real numbers R. Here AR is a compact topological group and a: R $\to$ AR is a continuous homomorphism with dense image. An important feature of the Bohr compactification is the following universal mapping property which it enjoys: given any compact topological group G and any continuous homomorphism ψ: R $\to$ G there exists a continuous homomorphism ϕ: AR $\to$ G such that $\psi = \phi \circ a$. Such universal mapping properties are central to the theory of semigroup compactifications.

Compactifications of semigroups can be produced in a variety of ways. One way is by the use of operator theory, a technique employed by deLeeuw and Glicksberg in their now classic 1961 paper on applications of almost periodic compactifications. In this setting, AR appears as the strong operator closure of the group of all translation operators on the C*-algebra AP(R) of almost periodic functions on R. More generally, but using essentially the same ideas, deLeeuw and Glicksberg were able to construct the almost periodic and weakly almost periodic compactifications of any semitopological semigroup with identity.

Another method of obtaining compactifications is based on the Adjoint Functor Theorem of category theory. The first systematic use of this technique appeared in the 1967 monograph of Berglund and Hofmann, where it was shown that any semitopological semigroup, with or without identity, possesses both almost periodic and weakly almost periodic compactifications. One important advantage of the category theory approach is that it provides a vantage point from which the fundamental unity of the subject may be viewed. In addition, category theory suggests other semigroup compactifications. An appendix here shows how the Adjoint Functor Theorem can be applied to produce a variety of semigroup compactifications.

A third method, and the one primarily used in this monograph, is based on the Gelfand-Naimark theory of commutative C*-algebras. Compactifications of a semitopological semigroup S now appear as the spectra of certain C*-algebras of functions on S. (For example, AR is taken as the spectrum of $\mathcal{AP}$(R).) This method yields (perhaps somewhat more elegantly) compactifications which could also be produced using the operator-theoretic approach, and still allows the use of functional analytic tools to facilitate their study. Furthermore it suggests a parallel theory of affine compactifications and provides a natural setting in which to study the interplay between the two theories via measure theory.

The main part of Chapter I is devoted to constructing compactifications (section 4). The necessary preliminary information about means on function spaces, from which the compactifications are constructed, is assembled in section 3. Sections 1 and 2 contain the basic facts and definitions concerning semigroups, flows, and probability measures on compact semigroups needed in later sections.

Chapter II is devoted primarily to structure theory. In section 1 the relevant algebraic structure theory is developed. The main result is the Rees-Suschkewitsch Theorem (Theorem 1.16). In the latter part of the section applications are made to transformation semigroups. Section 2 contains the structure theory of compact right topological semigroups. As might be expected, the theory is more complicated than the corresponding theory for compact semitopological (let alone topological) semigroups. One complication is the fact that, in contrast to the semitopological case, minimal right ideals and maximal subgroups of the minimal ideal need not be closed. The structures of compact right topological groups and of compact affine right topological semigroups are treated in sections 3 and 4 respectively. The last section of Chapter II examines the topologico-algebraic structure of the support of a mean on an algebra of functions defined on a semigroup.

Chapter III is the heart of the monograph. Much of the material presented in this chapter is new, beginning with the general theory of affine compactifications, the subject of section 1. The parallel theory of non-affine compactifications is treated in section 2. The emphasis of both of these sections is on the universal mapping property that a compactification enjoys (relative to the function space which defines the compactification). In sections 3-13, eleven different kinds of semigroup compactifications are constructed, including the familiar almost periodic and weakly almost periodic compactifications. The relevant functional analytic properties of the underlying function spaces are also examined. The universal mapping property that distinguishes each compactification is readily derived from the general theory developed in sections 1 and 2. General and specific inclusion relationships among the function spaces are presented in section 14; they suggest a dual theory of homomorphic image relationships among the corresponding compactifications. Section 15 treats the following interesting question: when can a function with certain properties on a subsemigroup S of a semigroup S' be extended to a function with the same properties on S'? This problem is essentially the same as the problem of determining when a compactification of S is canonically contained as a closed subsemigroup of the corresponding compactification of S'. The final section of Chapter III uses the structure theory developed in Chapter II to determine when a given C*-algebra of functions on a semigroup is a direct sum of an ideal of "flight functions" and a subalgebra of "reversible functions".

Chapter IV characterizes the existence of left invariant means on the function spaces of Chapter III in terms of the existence of fixed points for various types of flows. The presentation is in the spirit of the fixed point theorems of Day (1961) and Mitchell (1970).

Chapter V is a collection of examples which illuminate and test the sharpness of many of the results of previous chapters. It is by no means complete, a fact which we hope will inspire further research in the field.

The authors were influenced by many mathematicians before and during the preparation of this monograph. We would like to acknowledge our indebtedness, spiritual and otherwise, particularly to M. M. Day, I. Glicksberg, K. deLeeuw, J. S. Pym, K. H. Hofmann, T. Mitchell, and J. W. Baker.

Thanks go to Mrs. Wendy Waldie and Mrs. Barbara Smith for their skilful preparation of the typescript.

The research of the last-named author was partially supported by National Research Council of Canada grant A7857.

J. F. Berglund
H. D. Junghenn
P. Milnes

TABLE OF CONTENTS

CHAPTER I

PRELIMINARIES

1. SEMIGROUPS

1.1. Definition: A semigroup is a non-empty set S together with an associative binary operation $(s,t) \to st$: $S \times S \to S$, called multiplication. S is commutative if $st = ts$, $s, t \in S$.

If S is a semigroup, then for each $t \in S$ the maps

$$\rho_t: S \to S, \ \rho_t(s) = st$$

$$\lambda_t: S \to S, \ \lambda_t(s) = ts$$

are called, respectively, right and left multiplication maps (by t). We define

$$L(S) = \{\lambda_t \mid t \in S\}, \ R(S) = \{\rho_t \mid t \in S\}.$$

As a consequence of the identities

$$\lambda_{ts} = \lambda_t \lambda_s \ , \ \rho_{ts} = \rho_s \rho_t \ ,$$

the sets $L(S)$ and $R(S)$ are semigroups under composition of mappings.

If $s \in S$ and A, B are subsets of S, we shall write

$$sA = \lambda_s(A), \ As = \rho_s(A), \ AB = \bigcup_{s \in A} sB.$$

A non-empty subset T of a semigroup S is called a subsemigroup of S if $TT \subset T$, a right ideal of S if $TS \subset T$, or a left ideal of S if $ST \subset T$. A right (left) ideal of S which properly contains no right (left) ideal is called a minimal right (left) ideal.

An element $s \in S$ is called an idempotent if $s^2 = s$, a left (right) identity if $st = t$ $(ts = t)$ for all $t \in S$, or a left (right) zero if $st = s$ $(ts = s)$ for all $t \in S$. If s is both a left and a right identity (zero), it is simply called an identity (zero). A right (left) zero semigroup is one consisting entirely of right (left) zeros.

If S and T are semigroups, then $\psi: S \to T$ is a homomorphism if $\psi(st) = \psi(s)\psi(t)$ for all s, t $\in$ S.

1.2. Definition: Let S be a semigroup and a convex subset of a real or complex vector space E. S is an affine semigroup if the mappings ρ_t and λ_t are affine for all t $\in$ S.

1.3. Remark: If S is an affine semigroup, then, for any $a \geq 0$, $b \geq 0$ such that $a + b = 1$,

$$a\lambda_s + b\lambda_t = \lambda_{as+bt}, \quad s, t \in S.$$

Therefore L(S) is a convex subset of the vector space of all E-valued affine mappings on S. Furthermore,

$$(a\lambda_s + b\lambda_t)\lambda_r = a\lambda_s\lambda_r + b\lambda_t\lambda_r$$

and

$$\lambda_r(a\lambda_s + b\lambda_t) = a\lambda_r\lambda_s + b\lambda_r\lambda_t, \quad r, s, t \in S;$$

hence L(S) is an affine semigroup. Similarly, R(S) is an affine semigroup.

1.4. Definition: Let S be a semigroup with a Hausdorff topology. (All topologies considered here are assumed to be Hausdorff.) S is a right (left) topological semigroup if for each s $\in$ S the mapping ρ_s (λ_s) is continuous. If S is both left and right topological then S is called semitopological. Thus S is semitopological if and only if multiplication $(s,t) \to st: S \times S \to S$, is separately continuous. If this mapping is (jointly) continuous then S is a topological semigroup.

1.5. Example: Let X be a topological space and denote by X^X the set of all mappings from X into itself. If X^X has the product topology, then X^X is a right topological semigroup under function composition. Any subsemigroup of continuous mappings is a semitopological semigroup.

If X is a uniform space, then any subsemigroup S of equicontinuous mappings is a topological semigroup, as is the closure $\bar{S} \subset X^X$; and, if X is a compact uniform space, then the closure $\bar{S} \subset X^X$ of a subsemigroup S of equicontinuous mappings is a compact topological semigroup, and its (relativized) topology is that of uniform convergence on X [Kelley (1955); Ascoli Theorem, p. 233]. (See Proposition 2.2 (c) ahead for some details concerning the last statement.)

1.6. Definition: Let S be a semigroup. If f is a function on S we define

$$R_s f = f \circ \rho_s \text{ and } L_s f = f \circ \lambda_s \, , \; s \in S.$$

1.7. Remarks: If S is a semigroup, let $\mathcal{B}(S)$ denote the linear space of all bounded complex-valued functions on S. Then $\mathcal{B}(S)$ is a C*-algebra under the usual operations of (pointwise) addition and multiplication, under the uniform norm

$$\|f\| = \sup_{s \in S} |f(s)|,$$

and under involution $f \to f^*$, where $f^*(s) = \overline{f(s)}$, $s \in S$. The mappings R_s and L_s are *-homomorphisms of $\mathcal{B}(S)$ such that

$$R_{st} = R_s R_t, \; L_{st} = L_t L_s, \; s, \, t \in S.$$

If S is a right (left) topological semigroup, then, for

each $s \in S$, $R_s(L_s)$ maps $\mathcal{C}(S)$, the C*-subalgebra of all continuous functions in $\mathcal{B}(S)$, into itself.

Let F be a locally convex topological vector space with topological dual F^*. For a subset $\mathcal{D} \subset F^*$, we denote by $\sigma(F,\mathcal{D})$ the weak topology induced on F by $\mathcal{D}$.

1.8. Lemma: Let X and Y be topological spaces with X compact, and let $g: X \times Y \to C$ be a bounded function such that the function $x \to g(x,y)$ is continuous for each $y \in Y$.

(a) g is (jointly) continuous if and only if the mapping $y \to g(\cdot,y): Y \to \mathcal{C}(X)$ is norm continuous.

(b) If Y is compact, then g is separately continuous if and only if $y \to g(\cdot,y): Y \to \mathcal{C}(X)$ is $\sigma(\mathcal{C}(X),\mathcal{C}(X)^*)$ continuous.

Proof: (a) If $y \to g(\cdot,y)$ is not norm continuous, then there exist $\varepsilon > 0$ and a net $\{y_\alpha\} \subset Y$ converging to $y \in Y$ such that

$$\sup_{x \in X} |g(x,y_\alpha) - g(x,y)| \geq 2\varepsilon$$

for all α. For each α choose $x_\alpha \in X$ such that

$$|g(x_\alpha , y_\alpha) - g(x_\alpha , y)| \geq \varepsilon.$$

Since X is compact we may assume $\{x_\alpha\}$ converges to some $x \in X$. Then g cannot be jointly continuous; otherwise we would have $|g(x,y) - g(x,y)| \geq \varepsilon$.

Conversely, assume $y \to g(\cdot,y)$ is norm continuous, and let $\{x_\alpha\}$ be a net in X converging to $x \in S$, and $\{y_\beta\}$ a net in Y converging to $y \in Y$. Then from the inequality

$$\begin{aligned} &|g(x_\alpha , y_\beta) - g(x,y)| \\ &\leq |g(x_\alpha , y_\beta) - g(x_\alpha , y)| + |g(x_\alpha , y) - g(x,y)| \\ &\leq \|g(\cdot,y_\beta) - g(\cdot,y)\| + |g(x_\alpha , y) - g(x,y)|, \end{aligned}$$

it follows that $\{g(x_\alpha, y_\beta)\}$ converges to $g(x,y)$.

(b) Suppose Y is compact and g is separately continuous. Then $y \to g(\cdot,y)$ is continuous in the topology on $C(X)$ of pointwise convergence, and therefore $g(\cdot,Y)$ is compact in this topology; hence, on $g(\cdot,Y)$, this topology agrees with the topology $\sigma(C(X),C(X)^*)$ [Grothendieck (1952)]. Thus $y \to g(\cdot,y)$ is $\sigma(C(X),C(X)^*)$ continuous.

Conversely, if $y \to g(\cdot,y)$ is $\sigma(C(X),C(X)^*)$ continuous, then, in particular, $y \to g(x,y)$ is continuous for each $x \in X$, and g is separately continuous.

1.9. Corollary: Let S be a compact right topological semigroup.

(a) S is a topological semigroup if and only if

$$s \to R_s f\colon S \to C(S)$$

is norm continuous for each $f \in C(S)$. In this case, $s \to L_s f$ is norm continuous for each $f \in C(S)$.

(b) S is a semitopological semigroup if and only if

$$s \to R_s f\colon S \to C(S)$$

is $\sigma(C(S),C(S)^*)$ continuous for each $f \in C(S)$. In this case, $s \to L_s f$ is $\sigma(C(S),C(S)^*)$ continuous for each $f \in C(S)$.

Proof: (a) Since S is compact its topology is generated by $C(S)$. Therefore S is a topological semigroup if and only if $(s,t) \to f(st)\colon S \times S \to C$ is continuous for each $f \in C(S)$. By 1.8 (a) the latter condition is equivalent to the norm continuity of $t \to R_t f$.

(b) S is a semitopological semigroup if and only if

$(s,t) \to f(st): S \times S \to C$ is separately continuous for each $f \in C(S)$. By 1.8 (b), the latter condition is equivalent to the $\sigma(C(S),C(S)^*)$ continuity of $s \to R_s f$.

1.10. Definition: Let S be a compact semitopological semigroup, and let $C(S)^*$ denote the (topological) dual space of $C(S)$. Recall that by the Riesz Representation Theorem [Dunford and Schwartz (1964); Theorem IV.6.3], $C(S)^*$ may be identified with the space of all complex regular Borel measures on S, i.e., if $\mu \in C(S)^*$ and $f \in C(S)$, we may write

$$\mu(f) = \int f(s)\mu(ds).$$

By Corollary 1.9 (b), the mapping

$$t \to \int f(st)\mu(ds) = \mu(R_t f)$$

is continuous. Therefore, if μ, $\nu \in C(S)^*$, we may define

$$(1) \qquad (\mu * \nu)(f) = \iint f(st)\mu(ds)\nu(dt), \quad f \in C(S).$$

The linear functional $\mu * \nu$ is called the convolution of μ and ν. We will sometimes write $\mu\nu$ in place of $\mu * \nu$.

1.11. Remarks: It is easily seen that convolution is associative; hence $C(S)^*$ is a semigroup under convolution. Furthermore, if $e(s) \in C(S)^*$ is defined by

$$e(s)f = f(s), \quad f \in C(S), \quad s \in S,$$

then $e: S \to C(S)^*$ is an isomorphism and a homeomorphism onto the subsemigroup $e(S)$, where $C(S)^*$ has the $\sigma(C(S)^*,C(S))$ topology.

The next result shows that $C(S)^*$ is commutative if and only if S is commutative.

1.12. Lemma [Glicksberg (1961)]: Let S be a compact semitopological semigroup. Then for any $\mu, \nu \in C(S)^*$ and $f \in C(S)$,

$$(2) \qquad \iint f(st)\mu(ds)\nu(dt) = \iint f(st)\nu(dt)\mu(ds).$$

Proof: Let $Y = \{\eta \in C(S)^* \mid \|\eta\| \leq 1\}$. It is enough to prove that (2) holds for $\nu \in Y$. Clearly (2) is true if ν is a finite linear combination of members of $e(S)$. Since these are $\sigma(C(S)^*, C(S))$ dense in Y, (2) will follow if we show both sides are $\sigma(C(S)^*, C(S))$ continuous functions of ν. Now, the left side of (2) is a continuous function of ν by definition of the $\sigma(C(S)^*, C(S))$ topology. Also, 1.9 (b) implies that

$$(s,\nu) \to \int f(st)\nu(dt) = \nu(L_s f)$$

is separately continuous. Therefore

$$\nu \to \iint f(st)\nu(dt)\mu(ds)$$

is continuous by 1.8 (b).

1.13. A (regular Borel) probability measure on a compact semitopological semigroup S is a measure identified (via the Riesz Representation Theorem) with a functional $\mu \in C(S)^*$ satisfying $\mu(1) = \|\mu\| = 1$. We denote by P(S) the set of all such measures.

1.14. Theorem [Glicksberg (1959, 1961)]: Let S be a compact semitopological (topological) semigroup and let P(S) have the (relativized) $\sigma(C(S)^*, C(S))$ topology. Then P(S) is a compact affine semitopological (topological) semigroup under convolution.

Proof: P(S) is a closed subset of the unit ball of $C(S)^*$ and is therefore $\sigma(C(S)^*, C(S))$ compact. That P(S) is an affine left topological semigroup is clear from the definition of convolution, and equation (2) shows that P(S) is also right topological.

Now assume S is topological. If $\{\mu_\alpha\}$ is a net in P(S) which $\sigma(C(S)^*, C(S))$ converges to μ, then Lemma 1.8 implies that $\mu_\alpha(R_t f) \to \mu(R_t f)$ uniformly in $t \in S$ for each $f \in C(S)$. It follows that

$$(\mu_\alpha * \nu)(f) - (\mu * \nu)(f) = \int [\mu_\alpha(R_t f) - \mu(R_t f)]\nu(dt)$$

converges to zero uniformly in ν on bounded subsets of $C(S)^*$. Therefore, if $\{\nu_\beta\}$ is a net in P(S) which $\sigma(C(S)^*, C(S))$ converges to ν, then

$$\lim_{\alpha,\beta} (\mu_\alpha * \nu_\beta)(f) = (\mu * \nu)(f).$$

Thus P(S) is topological.

2. ACTIONS

2.1. Definitions: A transformation semigroup is a triple (X,S,π), where X is a non-empty set, S is a semigroup, and $\pi: S \times X \to X$, $(s,x) \to sx$, is a mapping which satisfies

$$(st)x = s(tx), \quad s, t \in S, \ x \in X.$$

The mapping π is called the action of the transformation semigroup and X is the phase space. When there is no danger of ambiguity the symbol π will be suppressed.

For each $s \in S$ define $\pi^s: X \to X$ by

$$\pi^s(x) = sx, \quad x \in X.$$

Note that $\pi^{st} = \pi^s \circ \pi^t$; hence, the set $\pi^S = \{\pi^s \mid s \in S\}$

is a subsemigroup of the semigroup X^X of all functions from X into itself (under the operation of composition of mappings), and $s \to \pi^s$ is a homomorphism.

A subset Y of the phase space X is said to be <u>invariant</u> under the action if $\pi(s,Y) \subset Y$ for all $s \in S$. In this case, $(Y,S,\pi|_{S\times Y})$ is a transformation semigroup. A member x of X is a <u>fixed point</u> of the action if $\{x\}$ is invariant, i.e., if $sx = x$ for all $s \in S$.

If X is a convex subset of a real or complex vector space, and if π^s is affine for each $s \in S$, then (X,S,π) is called an <u>affine transformation semigroup</u>.

A <u>flow</u> is a transformation semigroup (X,S,π) such that X is a compact topological space and π^s is continuous for each $s \in S$. The <u>enveloping semigroup</u> $E = E(X,S,\pi)$ of the flow (X,S,π) is the closure of π^S in the product space X^X. Since X^X is compact (by Tychonoff's Theorem) the enveloping semigroup is always compact. (Any mention of topology on X^X always refers to the usual product topology.) An affine transformation semigroup which is a flow is called an <u>affine flow</u>.

The next result will be useful in the sequel; see Definition IV.1.8 ahead.

2.2. <u>Proposition</u>: Let (X,S,π) be a flow with enveloping semigroup E, and let E have its relativized product topology.

(a) E is a right topological semigroup such that $\lambda_{\pi^s}: E \to E$ is continuous for each $s \in S$.

(b) If each member of E is continuous, then E is a semitopological semigroup.

(c) If π^B is equicontinuous (with respect to the unique uniformity on X) for a subset $B \subset S$, then the function

$$(\phi,\psi) \to \phi \circ \psi : (\pi^B)^- \times E \to E$$

is continuous, where $(\pi^B)^-$ is the closure of π^B in E.

(d) If π^S is equicontinuous, then E is a topological semigroup.

(e) If (X,S,π) is an affine flow then E is an affine semigroup.

Proof: To show that E is a semigroup, let $\phi, \psi \in E$. Then there exists nets $\{\pi^{s_\alpha}\}$ and $\{\pi^{t_\beta}\}$ converging pointwise to ϕ and ψ respectively. Since π^{s_α} is continuous,

$$\lim_\beta \pi^{s_\alpha t_\beta}(x) = \lim \pi^{s_\alpha}(\pi^{t_\beta}x) = \pi^{s_\alpha}(\psi(x))$$

for each $x \in X$, hence $\pi^{s_\alpha} \circ \psi \in E$. Taking limits with respect to α shows that $\phi \circ \psi \in E$. Parts (a) and (b) readily follow. (See Example 1.5.)

To prove (c) let $\phi_\alpha \to \phi$ in $\pi(B)^-$ and $\psi_\beta \to \psi$ in E. Note that, since π^B is equicontinuous, so is $(\pi^B)^-$ by Theorem 7.14, p. 232, of [Kelley (1955)]. Fix arbitrary $x \in X$ and a closed symmetric member U of the uniformity $\mathcal{U}$ of X. By the equicontinuity of $(\pi^B)^-$, there is a $V \in \mathcal{U}$ such that $(\psi(x),y) \in V$ implies that $(\psi' \circ \psi(x), \psi'(y)) \in U$ for every $\psi' \in (\pi^B)^-$. In particular,

$$(\phi_\alpha \circ \psi(x),\ \phi_\alpha(y)) \in U$$

for all α. Choose β_0 such that $(\psi(x), \psi_\beta(x)) \in V$ for all $\beta \geq \beta_0$. Then for such β and for all α,

$$(\phi_\alpha \circ \psi(x),\ \phi_\alpha \circ \psi_\beta(x) \in U.$$

Finally, choose α_0 such that for $\alpha \geq \alpha_0$,

$$(\phi_\alpha \circ \psi(x),\ \phi \circ \psi(x)) \in U.$$

Therefore $\alpha \geq \alpha_0$ and $\beta \geq \beta_0$ imply

$$(\phi \circ \psi(x),\ \phi_\alpha \circ \psi_\beta(x) \in U^2.$$

So $\lim_{\alpha,\beta} \phi_\alpha \circ \psi_\beta(x) = \phi \circ \psi(x)$, and we have (c).

Trivially, (d) follows from (c). And, if (X,S,π) is an affine flow, then, clearly, every member of E is affine; hence E is an affine semigroup, which proves (e).

2.3. <u>Definition</u>: A flow (X,S) is <u>distal</u> if $x = y$ whenever there exists a net $\{s_\alpha\}$ in S such that $\lim_\alpha s_\alpha x = \lim_\alpha s_\alpha y$.

2.4. <u>Remark</u>: A flow (X,S) is distal if and only if its enveloping semigroup is a group (not necessarily a topological group, however); see the end of Section II.3.

3. MEANS

Throughout this section, S denotes a topological space and F a conjugate closed, norm closed linear subspace of $C(S)$ containing the constant function 1.

3.1. Definition: A mean on F is a member μ of F^*, the dual space of F, such that $\mu(1) = 1 = \|\mu\|$. The set of all means on F is denoted by $M(F)$.

3.2. Proposition: Let $\mu \in M(F)$, $f \in F$.

(a) $\mu(f) \in$ closed convex hull in C of $f(S)$.

(b) $\mu(f) \in R$ if f is real valued.

(c) $\mu(f) \geq 0$ if $f \geq 0$.

(d) $\mu(f^*) = \overline{\mu(f)}$.

(e) $\mu(\mathrm{Re} f) = \mathrm{Re}\,\mu(f)$, $\mu(\mathrm{Im} f) = \mathrm{Im}\,\mu(f)$.

Conversely, if μ is a linear functional on F such that $\mu(1) = 1$ and $\mu(f) \geq 0$ for each $f \geq 0$, then μ is a mean on F.

Proof: The closed convex hull in C of $f(S)$ is the intersection of all closed disks in C containing $f(S)$. If D is such a disk, with center c and radius r, then since μ is a mean,

$$|\mu(f) - c| = |\mu(f - c)| \leq \|\mu\| \, \|f - c\| \leq r,$$

hence $\mu(f) \in D$. This proves (a). Parts (b) and (c) follow immediately from (a). Part (d) is a direct consequence of (b), and (e) in turn follows easily from (d).

Now suppose μ is a linear functional on F such that $\mu(1) = 1$ and $\mu(f) \geq 0$ for $f \geq 0$ in F. If $g \in F$ is real valued then $\|g\| - g \geq 0$ hence

$$\|g\| - \mu(g) = \mu(\|g\| - g) \geq 0.$$

Therefore $\mu(g)$ is real and $\mu(g) \leq \|g\|$. For arbitrary $f \in F$ choose $c \in C$ with $|c| = 1$ such that $|\mu(f)| = c\mu(f)$. Since F is conjugate closed there exist real-valued $g, h \in F$ such that $cf = g + ih$. Then

$$|\mu(f)| = \mu(cf) = \mu(g) + i\mu(h),$$

and since $\mu(g)$ and $\mu(h)$ are real, $\mu(h) = 0$ and

$$|\mu(f)| = \mu(g) \leq \|g\| \leq \|g + ih\| = |c| \ \|f\| = \|f\|.$$

Therefore $\mu \in F^*$ and $\|\mu\| = 1$.

3.3. Proposition: Every $\mu \in F^*$ is a finite linear combination of members of $M(F)$.

Proof: By the Hahn-Banach Theorem, μ may be extended to a continuous linear functional ν on $C(S)$. By the Gelfand-Naimark Theorem, $C(S)$ is isomorphic and isometric to $C(X)$ for some compact topological space X. By the Riesz Representation Theorem, ν may be interpreted as a measure on X and as such is a finite linear combination of probability measures on X (Jordan-Hahn Decomposition Theorem). These probability measures may be interpreted in the natural way as means on $C(S)$, and by restricting these means to F we obtain the desired representation of μ.

3.4. Definition: Abusing the notation established in 1.11, we define for each $s \in S$ a mean $e(s) \in M(F)$ by

$$e(s)f = f(s), \ f \in F.$$

The mean $e(s)$ is called evaluation at s, and $e: S \to M(F)$ is called the evaluation mapping.

A finite mean is any convex combination of members of $e(S)$. If μ is a finite mean then we may write

$$\mu = \sum_{s \in S} a(s)e(s),$$

where a: $S \to [0,1]$ is a function with finite support and $\sum_{s \in S} a(s) = 1$.

3.5. Proposition: $M(F)$ is convex and $\sigma(F^*,F)$ compact, and is the $\sigma(F^*,F)$ closure of the set of finite means on F. Furthermore, e: $S \to M(F)$ is $\sigma(F^*,F)$ continuous.

Proof: That $M(F)$ is convex and $\sigma(F^*,F)$ closed follows immediately from the alternate characterization of mean given in Proposition 3.2. Since $M(F)$ is contained in the unit ball of F^* it must be $\sigma(F^*,F)$ compact.

Now let $\mu \in M(F)$. If μ is not in the $\sigma(F^*,F)$ closure of the finite means, then by the Hahn-Banach Separation Theorem [Dunford and Schwartz (1964); Theorem V.2.10] there exists $f \in F$ such that

$$\mu(\mathrm{Ref}) > \sup_{\nu} \nu(\mathrm{Ref}), \tag{3}$$

where the supremum is taken over all means ν in the $\sigma(F^*,F)$ closure of the finite means. We may assume $\mathrm{Ref} \geq 0$ (otherwise replace f in (3) by $f + \|f\|$). Then

$$\sup_{\nu} \nu(\mathrm{Ref}) = \sup_{s \in S} e(s)\mathrm{Ref} = \|\mathrm{Ref}\|,$$

and we obtain the contradiction $\mu(\mathrm{Ref}) > \|\mathrm{Ref}\|$. Thus $M(F)$ is the $\sigma(F^*,F)$ closure of the finite means.

That e: $S \to M(F)$ is $\sigma(F^*,F)$ continuous follows from the fact that each $f \in F$ is continuous.

3.6. Lemma: Let X be a compact convex subset of a real locally convex topological vector space E, and $A_r(X)$ the Banach space of continuous real-valued affine functions on X.

If ξ is a mean on $A_r(X)$, then there exists $x \in X$ such that, for all $h \in A_r(X)$, $\xi(h) = h(x)$.

Proof: For each $h \in A_r(X)$ let

$$X_h = \{x \in X \mid h(x) = \xi(h)\}.$$

We must show that $\bigcap_{h \in A_r(X)} X_h \neq \emptyset$. Since X is compact it is enough to show that $\bigcap_{i=1}^{n} X_{h_i} \neq \emptyset$ for any finite collection $\{h_1,\ldots,h_n\} \subset A_r(X)$. Define a mapping $T: X \to R^n$ by $T(x) = (h_1(x),\ldots,h_n(x))$. Then T is affine and continuous, hence $T(X)$ is compact and convex. We must show that $(\xi(h_1),\ldots,\xi(h_n)) \in T(X)$. If not, then by the Hahn-Banach Separation Theorem there exists $(a_1,a_2,\ldots,a_n) \in R^n$ such that

$$\sum_{i=1}^{n} a_i\xi(h_i) > \sup_{x \in X} \sum_{i=1}^{n} a_ih_i(x).$$

Let $h = \sum_{i=1}^{n} a_ih_i$. Then $h \in A_r(X)$, and we have $\xi(h) > \|h\|$, contradicting that ξ is a mean.

3.7. <u>Corollary</u>: Let $A(M(F))$ denote the Banach space of all continuous complex-valued affine functions on $M(F)$, where $M(F)$ has the relativized $\sigma(F^*,F)$ topology. For each $f \in F$ define $\hat{f} \in A(M(F))$ by $\hat{f}(\mu) = \mu(f)$. Then $f \to \hat{f}$ is a linear isometry of F onto $A(M(F))$ such that $\hat{f} \circ e = f$.

Proof: Clearly $f \to \hat{f}$ is linear and

$$\|\hat{f}\| = \sup_{\mu \in M(F)} |\hat{f}(\mu)| = \sup_{s \in S} |\hat{f}(e(s))| = \|f\|.$$

If $\hat{F} \neq A(M(F))$, then since $\hat{F}$ is closed there exists a non-zero $\xi \in A(M(F))^*$ such that $\xi(\hat{F}) = \{0\}$. Assuming, as we may, that ξ is real, we may write $\xi = a_1\xi_1 - a_2\xi_2$, where $a_i \geq 0$

and ξ_i is a mean on $A(M(F))$, $i = 1, 2$ (proof of Proposition 3.3). By Lemma 3.6 there exist $\mu_i \in M(F)$ such that $\xi_i(h) = h(\mu_i)$ for all $h \in A(M(F))$, $i = 1, 2$. Hence, $a_1\mu_1(f) = a_2\mu_2(f)$ for all $f \in F$. But then $a_1 = a_2$, $\mu_1 = \mu_2$ and so $\xi = 0$. Therefore $f \to \hat{f}$ is surjective.

3.8. Definition: Let F be closed under (pointwise) multiplication. A mean μ on F is called multiplicative if

$$\mu(fg) = \mu(f)\mu(g), \quad f, g \in F.$$

We shall denote the set of all multiplicative means on F by $MM(F)$. Thus

$$MM(F) = \{\mu \in M(F) \mid \mu(fg) = \mu(f)\mu(g) \text{ for all } f, g \in F\}.$$

Note that $e(S) \subset MM(F)$.

If $F = C(S)$ we shall use the notation βS for $MM(F)$. If S is completely regular then βS is simply the Stone-Čech compactification of S.

3.9. Proposition: Let F be as in 3.8. Then $MM(F)$ is $\sigma(F^*,F)$ compact and is the $\sigma(F^*,F)$ closure of $e(S)$.

Proof: $MM(F)$ is obviously $\sigma(F^*,F)$ closed hence compact. By the Gelfand-Naimark Theorem, F is isomorphic and isometric to $C(MM(F))$ under $f \to \hat{f}$, where $\hat{f}(\mu) = \mu(f)$ and where $MM(F)$ has the $\sigma(F^*,F)$ topology. Since $\hat{f}(e(s)) = f(s)$, any function in $C(MM(F))$ which vanishes on $e(S)$ is identically zero. It follows from Urysohn's Lemma that $MM(F)$ must be the $\sigma(F^*,F)$ closure of $e(S)$.

3.10. Remark: Throughout the remainder of these notes, unless otherwise stated, any mention of topology on $M(F)$ or $MM(F)$ refers to the (relativized) $\sigma(F^*,F)$ topology.

4. SEMIGROUPS OF MEANS

Throughout this section S denotes a semitopological semigroup and F a conjugate closed, norm closed linear subspace of $C(S)$ containing the constant functions.

4.1. Definition: Let f be a function with domain S, and let $s \in S$. The functions

$$L_s f = f \circ \lambda_s, \quad R_s f = f \circ \rho_s$$

are called the left and right translates of f by s, respectively. The left and right orbits of f are the sets

$$L_S f = \{L_s f \mid s \in S\} \text{ and } R_S f = \{R_s f \mid s \in S\},$$

respectively.

4.2. Remark: L_s and R_s are bounded linear operators on $C(S)$ which satisfy

$$L_{st} = L_t L_s, \; R_{st} = R_s R_t, \quad s, t, \in S.$$

4.3. Definition: F is left (right) translation invariant if $L_s F \subset F$ $(R_s F \subset F)$ for all $s \in S$. F is translation invariant if it is both left and right translation invariant.

4.4. Remarks: If F is left translation invariant then L_s may be considered as an operator on F. Since $L_s f \geq 0$ whenever $f \in F$, $f \geq 0$, and since $L_s 1 = 1$, it follows from Proposition 3.2 that

$$L_s^* M(F) \subset M(F), \; s \in S,$$

where L_s^* denotes the adjoint of $L_s : F \to F$. Moreover, the mapping

$$(s,\mu) \to L_s^*\mu : S \times M(F) \to M(F)$$

is an affine action of S on $M(F)$, and relative to this

action $(M(F),S)$ is a flow. (As usual, $M(F)$ is assumed to have the relativized $\sigma(F^*,F)$ topology.)

If, in addition, F is a subalgebra, then $MM(F)$ is an invariant subset of $M(F)$ and hence, under the restricted action, $(MM(F),S)$ is a flow.

4.5. Definition: Let F be left (right) translation invariant. A mean μ on F is said to be left (right) invariant if, for each $f \in F$ and $s \in S$,

$$\mu(L_s f) = \mu(f) \quad (\mu(R_s f) = \mu(f)).$$

The set of all left (right) invariant means on F shall be denoted by $LIM(F)$ $(RIM(F))$. F is said to be left (right) amenable if $LIM(F) \neq \emptyset$ $(RIM(F) \neq \emptyset)$. If F is translation invariant and both left and right amenable, then F is said to be amenable.

4.6. Remarks: (a) The concept of amenability was introduced by Day (1957).

(b) If S is discrete and if $\mathcal{B}(S)$ is amenable, then S is said to be amenable. The reader is referred to Day's papers (1957, 1969) for results on amenable semigroups.

(c) If F is left translation invariant then $LIM(F)$ is simply the set of fixed points of the flow $(M(F),S)$:

$$LIM(F) = \{\mu \in M(F) \mid L_s^*\mu = \mu \text{ for all } s \in S\}.$$

4.7. Definition: If F is a subalgebra and is left (right) translation invariant, then F is extremely left (right) amenable if $MM(F) \cap LIM(F) \neq \emptyset$ $(MM(F) \cap RIM(F) \neq \emptyset)$. If F is translation invariant and both extremely left and extremely right amenable, then F is said to be extremely amenable.

The sets $MM(F) \cap LIM(F)$ and $MM(F) \cap RIM(F)$ shall be denoted by $MLIM(F)$ and $MRIM(F)$, respectively.

4.8. Remark: If S is discrete and if $B(S)$ is extremely amenable, then S is said to be extremely amenable. The reader is referred to [Granirer (1965, 1967)] for results on extremely amenable semigroups.

4.9. Definition: Let F be left translation invariant. For each $\nu \in F^*$ define $T_\nu : F \to B(S)$ by

$$(T_\nu f)(s) = \nu(L_s f), \quad f \in F, \ s \in S.$$

In the following lemma we list without proof some elementary properties of T_ν.

4.10. Lemma: Let F be left translation invariant and $\nu \in F^*$. Then T_ν has the following properties:

(a) T_ν is a bounded linear operator with $\|T_\nu\| \leq \|\nu\|$.

(b) $T_\nu L_s f = L_s T_\nu f$ for all $f \in F$, $s \in S$.

(c) $T_{e(s)} = R_s$ for all $s \in S$.

(d) If $\nu \in M(F)$ then T_ν is a positive operator, $T_\nu 1 = 1$, and $\|T_\nu\| = 1$.

(e) If F is an algebra and ν is a homomorphism then T_ν is a homomorphism.

(f) $\nu \to T_\nu$ is a linear mapping from F^* into the space of bounded linear transformations from F to $B(S)$.

4.11. Definition: Let F be left translation invariant. F is said to be left introverted if $T_\nu F \subset F$ for each $\nu \in M(F)$. If F is an algebra then F is said to be left m-introverted if $T_\nu F \subset F$ for each $\nu \in MM(F)$.

4.12. Remarks: (a) If F is left introverted or left m-introverted then F is right translation invariant (Lemma 4.10 (c)).

(b) An equivalent definition of left introversion is obtained by replacing the set $M(F)$ in Definition 4.11 by either $M(C(S))$ (since every mean on F extends to a mean on $C(S)$), or F^* (by virtue of Proposition 3.3). Similarly, in the definition of left m-introversion, $MM(F)$ may be replaced by βS.

(c) The notion of (left) introversion was introduced by Day (1957).

(d) Right introversion and right m-introversion are defined in an analogous manner. We shall rarely use these concepts, which would be involved in the dual theory of left topological semigroups.

4.13. Definition: Let F be left translation invariant. If F is left introverted (respectively, F is an algebra and is left m-introverted) and $\mu, \nu \in M(F)$ (respectively, $\mu, \nu \in MM(F)$), the product $\mu\nu \in F^*$ is defined by

$$\mu\nu = \mu \circ T_\nu .$$

4.14. Theorem: Let F be left translation invariant.

(a) If F if left introverted, then $(\mu,\nu) \to \mu\nu$ is a binary operation on $M(F)$ relative to which $M(F)$ is a compact affine right topological semigroup such that $\lambda_{e(s)}: M(F) \to M(F)$ is continuous for each $s \in S$.

(b) If F is a left m-introverted subalgebra, then $(\mu,\nu) \to \mu\nu$ is a binary operation on $MM(F)$ relative to which $MM(F)$ is a compact right topological semigroup such that

$\lambda_{e(s)}: MM(F) \to MM(F)$ is continuous for each $s \in S$.

In both (a) and (b) the evaluation mapping e is a homomorphism.

(c) If F is left introverted (left m-introverted) and left amenable (extremely left amenable), then $LIM(F) \neq \emptyset$ ($MLIM(F) \neq \emptyset$) and is a two-sided ideal of $M(F)$ ($MM(F)$).

Proof: (a) That $(\mu,\nu) \to \mu\nu$ is a binary operation on $M(F)$ follows from Lemma 4.10 (d) and Proposition 3.2. To verify associativity we show first that for $\mu, \nu \in M(F)$,

(4) $$T_{\mu\nu} = T_\mu \circ T_\nu.$$

Indeed, by Lemma 4.10 (b), if $f \in F$ and $s \in S$ then

$$(T_{\mu\nu}f)(s) = (\mu\nu)(L_s f) = \mu(T_\nu L_s f) = \mu(L_s T_\nu f)$$
$$= [T_\mu(T_\nu f)](s).$$

Thus, if $\lambda, \mu, \nu \in M(F)$, then

$$(\lambda\mu)\nu = (\lambda\mu) \circ T_\nu = (\lambda \circ T_\mu) \circ T_\nu = \lambda \circ (T_\mu \circ T_\nu)$$
$$= \lambda \circ (T_{\mu\nu}) = \lambda(\mu\nu).$$

Therefore $M(F)$ is a semigroup.

If $\mu \in M(F)$ then $\rho_\mu: M(F) \to M(F)$ is obviously affine and $\sigma(F^*,F)$ continuous. That $\lambda_\mu. M(F) \to M(F)$ is affine follows from Lemma 4.10 (f). The continuity of $\lambda_{e(s)}: M(F) \to M(F)$ follows from the observation that $\lambda_{e(s)} = L_s^*|_{M(F)}$.

To show that e: $S \to M(F)$ is a homomorphism we use Lemma 4.10 (c):

$$e(s)e(t) = e(s) \circ T_{e(t)} = e(s) \circ R_t = e(st).$$

Part (b) is proved similarly, and (c) follows from Remark 4.15 (b), ahead.

4.15. Remarks: (a) With respect to the semigroup $M(F)$,

$$\lambda_{e(s)} = L_s^*|_{M(F)} \text{ and } \rho_{e(s)} = R_s^*|_{M(F)}.$$

A similar remark holds for $MM(F)$ (when F is an algebra).

(b) If F is left introverted and left amenable (i.e., $LIM(F) \neq \emptyset$), then $LIM(F)$ is the set of all right zeros of the semigroup $M(F)$, and hence is a right zero semigroup and a two-sided ideal of $M(F)$. Indeed, if ν is a right zero, then by the previous remark $L_s^*\nu = \lambda_{e(s)}(\nu) = \nu$. Conversely, if $\nu \in LIM(F)$, then for any $s \in S$, $\rho_\nu(e(s)) = \lambda_{e(s)}(\nu) = L_s^*\nu = \nu$, and since ρ_ν is affine and $\sigma(F^*,F)$ continuous it follows from Proposition 3.5 that $\mu\nu = \rho_\nu(\mu) = \nu$ for all $\mu \in M(F)$.

Similarly, if F is a left m-introverted subalgebra and is extremely left amenable, then $MLIM(F)$ is the set of all right zeros of $MM(F)$ and is a two-sided ideal of $MM(F)$.

(c) Equation (4) shows that for left introverted F, the mapping $\mu \to T_\mu$ is a representation of the semigroup $M(F)$ by bounded linear operators on F. If S has an identity (or, more generally, if there exists an $s \in S$ such that $L_sF = F$), then the mapping is one-to-one.

Similar remarks apply in the left m-introverted case.

(d) If F is left introverted then the multiplication $(\mu,\nu) \to \mu\nu = \mu \circ T_\nu$ is defined for <u>any</u> pair of continuous linear functionals μ, ν on F. (See Remark 4.12 (b).) With respect to this multiplication and ordinary addition and scalar multiplication, F^* is easily seen to be a Banach algebra. Equation (4) then holds for all $\mu, \nu \in F^*$; hence, in view of Lemma 4.10 (a,f), $\mu \to T_\mu$ is a norm continuous representation of the algebra F^* by bounded linear operators on F.

(e) The linear functional $\mu\nu$ is called the <u>evolution product</u> of μ and ν in [Pym (1964)]. (In Pym's notation, $\mu\nu = \mu \circ \nu$.) In the same paper, for right introverted F, the <u>convolution product</u> $\mu \# \nu$ is also defined. Let $U_\mu: F \to F$ be the operator defined by

$$(U_\mu f)(s) = \mu(R_s f), \quad f \in F, \; s \in S.$$

Then $\mu \# \nu$ ($\mu * \nu$ in Pym's notation) is the linear functional

$$\mu \# \nu = \nu \circ U_\mu .$$

F^* with convolution product is easily seen to be a Banach algebra. In fact, it is shown in [Pym (1969)] that F^* is a quotient Banach algebra of A^{**}, where A is some Banach algebra and A^{**} has Arens multiplication. (See [Pym (1969)] for details and references.)

If F is both left and right introverted, then the two multiplications on F^* are identical if and only if the convolution (or evolution) product is $\sigma(F^*,F)$ separately continuous [Pym (1964); Theorem 5.2].

4.16. <u>Theorem</u>: Let F be a translation invariant left m-introverted C*-subalgebra of $C(S)$ such that $1 \in F$. Let $X = MM(F)$ furnished with the $\sigma(F^*,F)$ topology, and let $f \to \hat{f}: F \to C(X)$ be the Gelfand mapping defined by $\hat{f}(x) = x(f)$. For each $\mu \in M(F)$ define the probability measure $\hat{\mu}$ on X by

$$\mu(f) = \int_X \hat{f}(x)\hat{\mu}(dx), \quad f \in F.$$

If X is a semitopological (topological) semigroup with respect to multiplication defined in 4.13 then F is left introverted, $M(F)$ is a semitopological (topological) semigroup with respect to multiplication defined in 4.13, and $\mu \to \hat{\mu}: M(F) \to P(X)$ is an isomorphism of affine semigroups, where $P(X)$ denotes the semigroup of regular Borel probability measures on X under convolution (Theorem 1.14).

Proof: For each $\nu \in M(F)$, $f \in F$, the function $y \to \int_X \hat{f}(yx)\hat{\nu}(dx)$ is continuous by Lemma 1.8. Since F is isometrically isomorphic (via the mapping $h \to \hat{h}$) to $C(X)$, there exists $g \in F$ such that

$$\hat{g}(y) = \int_X \hat{f}(yx)\hat{\nu}(dx), \quad y \in X.$$

In particular,

$$g(s) = \hat{g}(e(s)) = \int_X \hat{f}(e(s)x)\hat{\nu}(dx), \quad s \in S,$$

where $e\colon S \to X$ is the evaluation map. But

$$\hat{f}(e(s)x) = \hat{f}(L_s^* x) = (L_s^* x)(f) = x(L_s f) = (L_s f)^{\wedge}(x).$$

Therefore $g(s) = \int_X (L_s f)^{\wedge}(x)\hat{\nu}(dx) = \nu(L_s f)$, $s \in S$; hence F is left introverted.

Now let $\mu, \nu \in M(F)$, and let $T_\nu\colon F \to F$ be the operator defined in 4.9. For any $f \in F$, $s \in S$,

$$(T_\nu f)^{\wedge}(e(s)) = (T_\nu f)(s) = \nu(L_s f) = \int_X (L_s f)^{\wedge}(x)\hat{\nu}(dx)$$

$$= \int_X \hat{f}(e(s)x)\hat{\nu}(dx),$$

and it follows by continuity and Proposition 3.9 that

$$(T_\nu f)^{\wedge}(y) = \int_X \hat{f}(yx)\hat{\nu}(dx), \quad y \in X.$$

Therefore $(\mu\nu)(f) = \mu(T_\nu f) = \int_X (T_\nu f)^{\wedge}(y)\hat{\mu}(dy)$

$$= \int_X \int_X \hat{f}(yx)\hat{\nu}(dx)\hat{\mu}(dy);$$

hence $(\mu\nu)^{\wedge} = \hat{\mu} * \hat{\nu}$. Thus $\mu \to \hat{\mu}$ is a homomorphism. It is clearly an affine homeomorphism (relative to the $\sigma(F^*,F)$ and $\sigma(C(X)^*,C(X))$ topologies). That $M(F)$ is semitopological (topological) now follows from 1.14.

The remainder of this section consists of some technical results which shall be of use in later chapters.

4.17. Lemma: Let S be a semitopological semigroup and let τ be a locally convex topology on $C(S)$ such that $p \leq \tau \leq u$, where p denotes the topology of pointwise convergence on $C(S)$ and u the uniform (norm) topology. Suppose the maps, $f \to f^*: C(S) \to C(S)$ and, for each $t \in S$, $L_t: C(S) \to C(S)$, are τ-continuous and let

$$F = \{f \in C(S) \mid coR_S f \text{ is relatively } \tau\text{-compact}\},$$

where $coR_S f$ denotes the convex hull of $R_S f$. Then F is a u-closed, conjugate closed, translation invariant, left introverted subspace of $C(S)$ containing the constant functions.

Proof: The relations

$$coR_S(af + bg) \subset a\, coR_S f + b\, coR_S g$$
$$coR_S f^* = (coR_S f)^*$$
$$coR_S L_t f = L_t coR_S f$$
$$coR_S R_t f \subset coR_S f,$$

together with the hypotheses, show that F is a conjugate closed, translation invariant, linear subspace of $C(S)$. To show that F is u-closed, let $\{f_n\}$ be a sequence in F and $f \in C(S)$ such that $\|f_n - f\| \to 0$. Let $\{\sum_{s\in S} a_\alpha(s)R_s f\}$ be a net in $coR_S f$; here $a_\alpha: S \to [0,1]$ has finite support and

$\sum_{s \in S} a_\alpha(s) = 1$. If $e: S \to M(C(S))$ denotes the evaluation mapping, then $\{\sum_{s \in S} a_\alpha(s)e(s)\}$ has a subnet $\{\sum a_\beta(s)e(s)\}$ which $\sigma(C(S)^*, C(S))$ converges to some mean μ on $C(S)$. For each n and each $t \in S$ set

$$g_n(t) = \mu(L_t f_n), \quad g(t) = \mu(L_t f).$$

Then $g_n(t) = \lim_\beta \sum_{s \in S} a_\beta(s)e(s)(L_t f_n) = \lim_\beta \sum_{s \in S} a_\beta(s)R_s f_n(t)$, and, since $\text{co}R_S f_n$ is relatively τ-compact and $\tau \geq p$, $\{\sum_{s \in S} a_\beta(s)R_s f_n\}$ must τ-converge to g_n. Thus $g_n \in C(S)$ and, since $\{g_n\}$ converges uniformly on S to g, $g \in C(S)$. To show that $f \in F$ it now suffices to show that $\{\sum_{s \in S} a_\beta(s)R_s f\}$ τ-converges to g. Let q be a τ-continuous seminorm on $C(S)$ and $\varepsilon > 0$. Since $\tau \leq u$, there exists $c > 0$ such that

$$q(h) \leq c\|h\|, \quad h \in C(S). \tag{5}$$

Now choose n such that $\|f_n - f\| < \varepsilon$, and note that this implies $\|g_n - g\| < \varepsilon$. Hence, if β_0 is chosen so that

$$q(\sum a_\beta(s)R_s f_n - g_n) < \varepsilon, \ \beta \geq \beta_0,$$

then the triangle inequality and (5) imply that

$$q(\sum a_\beta(s)R_s f - g) < (1 + 2c)\varepsilon, \ \beta \geq \beta_0,$$

as required. Therefore $f \in F$, and F is u-closed.

To show that F is left introverted, let $f \in F$, $\mu \in M(C(S))$, and set $g(t) = \mu(L_t f)$. Let $\{\sum_{s \in S} a_\alpha(s)e(s)\}$ be a net of finite means on $C(S)$ which $\sigma(C(S)^*, C(S))$ converges to μ. Then, arguing as in the preceding paragraph,

$\{\sum_{s\in S} a_\alpha(s)R_s f\}$ converges to g pointwise, and $g \in C(S)$. If $\sum_{t\in S} b(t)R_t g$ is any member of $\mathrm{co}R_S g$, then, for each $r \in S$, $\sum_{t\in S} b(t)R_t g(r) = \lim_\alpha \sum_{s,t\in S} b(t)a_\alpha(s)R_{ts}f(r)$. Since $\sum_{s,t\in S} b(t)a_\alpha(s)R_{ts}f \in \mathrm{co}R_S f$ and $f \in F$, it follows that $\sum_{t\in S} b(t)R_t g$ is in the τ-closure of the relatively τ-compact set $\mathrm{co}R_S f$. Therefore $g \in F$.

4.18. Lemma: Let S and τ be as in Lemma 4.17, and let

$$F = \{f \in C(S) \mid R_S f \text{ is relatively } \tau\text{-compact}\}.$$

Then F is a u-closed, conjugate closed, translation invariant subspace of $C(S)$ containing the constant functions. Furthermore, if F is closed under (pointwise) multiplication then F is left m-introverted.

Proof: Analogous to that of 4.17.

The proof of the following lemma is straightforward and is left to the reader.

4.19. Lemma: Let F be a translation invariant, conjugate closed, norm closed, linear subspace of $C(S)$ containing the constant functions, and let $f \in F$.

(a) If F is left introverted, then the p-closure of $\mathrm{co}R_S f$ is $\{T_\mu f \mid \mu \in M(F)\}$.

(b) If F is a left m-introverted subalgebra of $C(S)$, then the p-closure of $R_S f$ is $\{T_\mu f \mid \mu \in MM(F)\}$.

CHAPTER II

THE STRUCTURE OF COMPACT SEMIGROUPS

In this chapter we investigate the structure of compact right topological semigroups. The existence of an idempotent in such a semigroup (Proposition 2.1) leads to the structure theorem for the minimal ideal (Theorem 2.2). This theorem is quite satisfactory algebraically, but, for example, minimal right ideals need not be closed or pairwise topologically isomorphic, neither of which can occur in the semitopological setting. In Section 3 we exhibit two different kinds of compact right topological groups. Sections 4 and 5 deal with compact affine right topological semigroups and the support of means, respectively. Further assertions which hold in the semitopological setting are seen to fail in the right topological setting, and some results are pushed through.

1. ALGEBRA

This section contains some information of a purely algebraic nature. The first part of it leads up to the statement of the structure theorem for a completely simple minimal ideal (Theorem 1.16). The last part is inspired by some results of deLeeuw and Glicksberg (1961); and parts of our Theorems 1.30, 1.34 and 1.37 correspond to parts of their Theorems 4.10, 4.9 and 7.4, respectively. Whereas their results are about bounded semigroups of linear operators on a Banach space, ours are about semigroups of transformations of a set X. In the intermediate setting of semigroups of operators on a vector space, some results of this kind were also developed in [Berglund and Hofmann (1967); II.1].

1.1. Definitions: Let S be a semigroup and let A be a non-empty subset of S. We recall that A is

(1) a subsemigroup if $AA = \{st \mid s, t \in A\} \subset A$;

(2) a right ideal if $AS \subset A$;

(3) a left ideal if $SA \subset A$;

(4) an ideal if it is both a left and a right ideal;

and (5) a minimal (left) [right] ideal if it contains no proper (left) [right] ideals.

The semigroup S is called

(1) left simple if it contains no proper left ideals;

(2) right simple if it contains no proper right ideals;

and (3) simple if it contains no proper ideals.

The semigroup S is called

(1) left cancellative if, for every $x, y \in S$, there is an $s \in S$ with

$$sx = sy$$

only if $x = y$;

(2) right cancellative if, for every $x, y \in S$, there is an $s \in S$ with

$$xs = ys$$

only if $x = y$; and

(3) cancellative if it is both left cancellative and right cancellative.

Finally, call S commutative if $st = ts$ for every $s, t \in S$.

1.2. Remark: Ordinarily one notes that for every "left" statement, there is a corresponding "right" statement, and, therefore, time and effort, as well as paper, are conserved by making only "right" statements, say, while leaving the "left" statements to the imagination of the reader. Having fixed our attention in the rest of these notes, however, on a particular one-sided topological notion, we think it will be clearer in the sequel if we keep the algebraic "left" and "right" somewhat separate.

1.3. Definitions: Let S be a semigroup and let $e \in S$. We recall that e is

(1) an idempotent if $e^2 = e$;

(2) a right identity if $se = s$ for every $s \in S$;

(3) a left identity if $es = s$ for every $s \in S$;

(4) an identity if it is both a left identity and a right identity;

(5) a left zero if $es = e$ for every $s \in S$;

(6) a right zero if $se = e$ for every $s \in S$; and

(7) a zero if it is both a left zero and a right zero.

1.4. Definitions: Let S be a semigroup. We call S

(1) a right zero semigroup if $st = t$ for every $s, t \in S$;

(2) a left zero semigroup if $st = s$ for every $s, t \in S$.

For a semigroup S, we let E(S) denote the set of idempotents in S. There is a natural partial order $\leq$ on E(S) given by

$$e \leq f \text{ if and only if } ef = fe = e.$$

An idempotent which is minimal with respect to this partial order is called <u>primitive</u>. A simple semigroup containing a primitive idempotent is <u>completely simple</u>.

1.5. <u>Lemma</u>: Let S be a semigroup, and let $e \in E(S)$.

(a) If S is right [left] simple then e is a left [right] identity.

(b) If S is completely simple and e is primitive, then eS [Se] is a minimal right [left] ideal.

Proof: (a) is obvious. For (b), suppose R is a right ideal contained in eS and $s \in R$. Then $s = es$, and $S = SsS$, since S is simple; hence $e = rst$ for some $r, t \in S$ and we may assume $r = ere$, $t = te$. Then $f = str = f^2$, $ef = f = fe$, and $f = e$, hence $R \supset sS \supset strS = eS$, as required.

1.6. <u>Theorem</u>: Let S be a semigroup. The following statements are equivalent:

(a) S is cancellative, simple, and contains an idempotent.

(b) S is left simple and right simple.

(c) For every $a, b \in S$, the equations

$$ax = b \text{ and } ya = b$$

are solvable in S.

(d) S has a right identity e and for every element $t \in S$ the equation

$$tx = e$$

is solvable in S.

(e) S is a group.

Proof: (a) implies (b). Let e be an idempotent in S. By cancellation, e is an identity for S. Suppose $A \subset S$ is a left ideal. Since S is simple, AS = S. Thus, for some $a \in A$ and some $s \in S$, we have e = as. Then

$$sas = se = es$$

and, canceling on the right, we get

$$e = sa \in A.$$

Thus, $S = Se \subset A$, and we conclude that S is left simple. Likewise, S is right simple.

That (b) implies (c) is clear.

(c) implies (d). Fix $a \in S$. Let x = e be the solution to ax = a. Then if $b \in S$ is arbitrary and y is such that ya = b, we have

$$be = yae = ya = b.$$

Thus e is a right identity.

(d) implies (e). Solving tx = e, then xy = e, we get t = te = t(xy) = (tx)y = ey = e(ey) = et. Hence e is an identity; also t = y, tx = xt = e and S is a group.

That (e) implies (a) (or any of the others) is obvious.

1.7. <u>Theorem</u>: The following assertions concerning a semigroup S are equivalent:

(a) S is right simple and left cancellative.

(b) S is right simple and contains an idempotent.

(c) S has a left identity e and for every element $t \in S$, the equation

$$tx = e$$

is solvable in S.

(d) S is the direct product $G \times Y$ of a group G and a right zero semigroup Y.

Specifically, under these circumstances, if $e \in E(S)$ is fixed, then

(i) $G = Se$ is a group,

(ii) $Y = E(S)$ is a right zero semigroup, and

(iii) the function $\zeta_e: G \times Y \to S$ defined by

$$\zeta_e(g,y) = gy$$

is an isomorphism with inverse

$$\zeta_e^{-1}(s) = (se,(se)^{-1}s),$$

where $(se)^{-1}$ is the inverse of se in the group G.

Also, each idempotent $e \in S$ is primitive.

Proof: (a) implies (b). Fix $a \in S$. Since S is right simple, $aS = S$. Let $x = e$ be a solution to $ax = a$. Then $ae = (ae)e = ae^2$; and, canceling a, we have $e = e^2$.

(b) implies (c). By Lemma 1.5, e is a left identity. Since S is right simple, $tS = S$; thus, there is some x so that $tx = e$.

(c) implies (d). Let $f \in E(S)$. Since $fx = e$ for some $x \in S$, we have $fe = f(fx) = fx = e$. Thus f is also a left identity and $Y = E(S)$ is a right zero semigroup.

Since e is idempotent, it follows that, for any $t \in S$, the equation $tx = e$ has a solution in Se. Since e is a right identity for $G = Se$, we get from Theorem 1.6 (d) that G is a group. Now,

$$\begin{aligned}[\zeta_e(g_1,y_1)][\zeta_e(g_2,y_2)] &= (g_1y_1)(g_2y_2) = g_1(y_1g_2)y_2 \\ &= g_1g_2y_2 = (g_1g_2)(y_1y_2) \\ &= \zeta_e[(g_1,y_1)(g_2,y_2)].\end{aligned}$$

And $\zeta_e(\zeta_e^{-1}(s)) = (se)[(se)^{-1}s)] = es = s$. Also

$$\zeta_e^{-1}(\zeta_e(g,y)) = (gye,(gye)^{-1}(gy)) = (g,g^{-1}(gy))$$
$$= (g,ey) = (g,y).$$

(d) implies (a). Right simplicity and left cancellativity are preserved under products.

The last assertion of the theorem is easily verified.

1.8. <u>Definition</u>: A semigroup S satisfying the equivalent conditions of Theorem 1.7 is called a <u>right-group</u>. A semigroup satisfying the equivalent conditions obtained by interchanging the words "right" and "left" in Theorem 1.7 is a <u>left-group</u>.

1.9. <u>Lemma</u>: If R is a minimal right ideal of a semigroup S, then $R = tS$ for any element $t \in R$.

1.10. <u>Theorem</u>: If S is a semigroup and $R \subset S$ is a minimal right ideal, then R is a right-group if and only if $E(R)$ is not empty.

Proof: From Lemma 1.9 we get that R is right simple, and the result follows from Theorem 1.7.

1.11. <u>Lemma</u>: If R is a minimal right ideal of a semigroup S, and I is an ideal of S, then $R \subset I$.

1.12. <u>Lemma</u>: If R is a minimal right ideal of a semigroup S and if $a \in S$, then aR is a minimal right ideal of S. Moreover, every minimal right ideal R_0 of S is given by $R_0 = tR$ for some $t \in S$.

Proof: If R' is a right ideal of S which is contained in aR, then

$$A = \{s \in R \mid as \in R'\}$$

is a right ideal of S contained in R; hence, $A = R$. But, clearly, $R' = aA$; so, $R' = aR$, as desired.

Now, if R_0 is any minimal right ideal of S, and $t \in R_0$, then tR is a right ideal of S contained in R_0. Thus, $R_0 = tR$.

1.13. Theorem: Let e be an idempotent in a semigroup S. Then the following statements are equivalent:

(a) eS is a minimal right ideal of S.

(b) eS is a right-group.

(c) eS is right simple.

(d) eSe is the maximal subgroup of S containing e.

(e) Se is a minimal left ideal of S.

(f) Se is a left-group.

(g) Se is left simple.

(h) e is primitive.

Proof: We prove the equivalence of (a), (b), (c), (d) and (h). The rest of the equivalences will follow from left-right duality since (d) and (h) are not one-sided statements.

That (a) implies (b) follows from Theorem 1.10, while Theorem 1.7 shows the equivalence of (b) and (c) and that (c) implies (d). Obviously (c) implies (a) and (b) implies (h); that (h) implies (a) was proved in Lemma 1.5.

(d) implies (b). eS has a left identity e, and if $t \in eS$, then te is in the group eSe. Thus, there is an inverse $(te)^{-1}$ for te in eSe. Now, if $x = (te)^{-1}$, we have

$tx = t(te)^{-1} = t[e(te)^{-1}] = (te)(te)^{-1} = e$; and, therefore, eS is a right-group by Theorem 1.7.

1.14. Lemma: If the semigroup S contains a minimal right ideal, then it contains a minimal ideal K(S), which is the union of all the minimal right ideals of S.

Proof: After Lemma 1.11, we need only show that the set

$$K = K(S) = \cup \{R \mid R \text{ is a minimal right ideal}\}$$

is an ideal. Clearly, it is a right ideal. Let $m \in K$ and $s \in S$. Let R be a minimal right ideal such that $m \in R$. Then mR is a minimal right ideal by Lemma 1.12. Thus

$$sm \in sR \subset K;$$

and we conclude that K is a left ideal.

1.15. Remarks: Most of the following theorem now follows readily from the results of this section. (The reader who wishes more detail can refer to [Berglund and Hofmann (1967); especially II.1], for example.) We remark that the structure theorem we get in the next section (Theorem 2.2) for the minimal ideal of a compact right topological semigroup is quite satisfactory algebraically; its "failings" are of a topological nature.

1.16. Theorem: Suppose that S contains a minimal ideal K(S). Then the equivalent statements of Theorem 1.13 about an idempotent $e \in S$ imply the following equivalent statements:

(i) $e \in K(S)$;

(j) $K(S) = SeS$.

If conditions (a) - (j) hold for some idempotent e, then e

is a primitive idempotent and $K(S)$ is completely simple. Conversely, if $K(S)$ contains a primitive idempotent (equivalently, if $K(S)$ has an idempotent that is contained in a minimal left or right ideal), then every minimal left ideal L [minimal right ideal R] is of the form $L = Sf$ [$R = fS$] for some $f \in E(K(S))$; and, furthermore,

$$\begin{aligned} K(S) &= \cup \{Sf \mid f \in E(K(S))\} \\ &= \cup \{fS \mid f \in E(K(S))\} \\ &= \cup \{fSf \mid f \in E(K(S))\}. \end{aligned}$$

Also, if $K(S)$ contains a primitive idempotent e, then $K(S)$ is isomorphic to $E(eS) \times eSe \times E(Se)$, which has multiplication

$$(u,v,w)(x,y,z) = (u,vwxy,z).$$

The isomorphism is given by

$$s \to (s(ese)^{-1}, ese, (ese)^{-1}s)$$

(inversion being in the group eSe) and its inverse is $(u,v,w) \to uvw$. If f is any other idempotent in $K(S)$, then the map $t \to ftf$ is an isomorphism of eSe onto fSf.

1.17. <u>Theorem</u>: Suppose the semigroup S contains a completely simple minimal ideal $K(S)$. Fix $e \in E(K(S))$, and let R be the minimal right ideal eS. Then the following statements about an element $s \in S$ are equivalent:

(a) $se \in eS$.

(b) $se = ese$.

(c) $sR \subset R$.

(d) $sf \in fS$ for every $f \in E(R)$.

Proof: Straightforward.

1.18. <u>Definition</u>: If X is a set, X^X denotes the set of all self-maps of X. X^X is a semigroup under composition and, for the rest of this section S will be a subsemigroup of X^X. If S is restricted to an invariant subset Y of X, $S|_Y$ will be considered as a subset of Y^Y (i.e., $S|_Y$ is a homomorphic image of S).

1.19. <u>Theorem</u>: Suppose $S \subset X^X$ is a subsemigroup which is a left zero semigroup. Then for $s, t \in S$, we have $sX = tX$ if and only if $s = t$.

Proof: Suppose $sX = tX$. Fix $x \in X$. Then there is some $x' \in X$ with $sx = tx'$. Since S is left zero, we have

$$sx = tx' = t^2x' = ttx' = tsx = tx.$$

Since x was arbitrary, we have $s = t$. The converse is trivial.

1.20. <u>Lemma</u>: Suppose $S \subset X^X$ is right simple (e.g., S is a right-group). Then $sX = tX$ for all $s, t \in S$.

1.21. <u>Definition</u>: Suppose $S \subset X^X$. Each $s \in S$ induces an equivalence relation $\sim$ on X defined by

$$x \sim y \text{ if and only if } sx = sy.$$

The partition of X into equivalence classes is what interests us most. We will speak therefore of <u>the partition of X by s</u>.

1.22. <u>Theorem</u>: Suppose $s, t \in S \subset X^X$.

(i) If S is a left zero semigroup, then the partition of X by s is the same as the partition of X by t.

(ii) If S is a right zero semigroup, then the partition of X by s is the same as the partition of X by t

if and only if $s = t$.

Proof: (i) Suppose $sx = sy$. Then, if S is a left zero semigroup

$$tx = (ts)x = t(sx) = t(sy) = ty.$$

(ii) Suppose for $x, y \in X$ we have $sx = sy$ if and only if $tx = ty$. Fix $x \in X$. Since S is right zero, Lemma 1.20 implies that there is some $z \in X$ with $sx = tz$. But then $sx = tz = stz$; so

$$tx = ttz = tsx = sx.$$

Since x was arbitrary, we have $t = s$, as desired. The converse is trivial.

1.23. Theorem: Suppose $S \subset X^X$ has a completely simple minimal ideal $K(S)$. Fix $e \in E(K(S))$; and let R be the minimal right ideal eS. Then the following statements about an element $s \in S$ are equivalent:

(a) $se \in eS$.

(b) $se = ese$.

(c) $sR \subset R$.

(d) $sf \in fS$ for every $f \in E(R)$.

(e) $seX = eX$.

Proof: The equivalence of (a) - (d) was established in Theorem 1.17, and Lemma 1.20 shows that (a) implies (e).

To establish (e) implies (b), fix $x \in X$, and choose x' so that $sex = ex'$. Then

$$esex = e^2x' = ex' = sex.$$

Since this holds for every $x \in X$, we have $ese = se$.

1.24. Theorem: Suppose $S \subset X^X$ has a completely simple minimal ideal $K(S)$. Fix $e \in E(K(S))$; and let L be the minimal left ideal Se. Then the following statements about an element $s \in S$ are equivalent:

(a) $es \in Se$.

(b) $es = ese$.

(c) $Ls \subset L$.

(d) $fs \in Sf$ for every $f \in E(L)$.

(e) The partition of X by e is the same as the partition of X by es.

Proof: The equivalence of (a) - (d) is the dual to Theorem 1.17.

Suppose that for some $x, y \in X$, we have $esx = esy$. Now, if $es = ese$, then

$$\begin{aligned} ex &= (ese)^{-1}(ese)x = (ese)^{-1}esx \\ &= (ese)^{-1}esy = (ese)^{-1}esey = ey. \end{aligned}$$

On the other hand, if $ex = ey$, then

$$esx = esex = esey = esy.$$

Thus (b) implies (e). For the converse, note that for each $x \in X$, $ex = e(ex)$. Thus, $esx = es(ex)$ for every x and, therefore, $es = ese$.

1.25. Lemma: Suppose the semigroup S has a completely simple minimal ideal $K(S)$. For a minimal right ideal R, define

$$S_R = \{s \in S \mid sR \subset R\}.$$

Then S_R is a subsemigroup of S with minimal ideal

$$K(S_R) = R.$$

Proof: [Berglund and Hofmann (1967); II.1.20].

1.26. Definition: Suppose that $S \subset X^X$ has a completely simple minimal ideal $K(S)$, and that $e \in E(K(S))$. Let L be the minimal left ideal Se and let R be the minimal right ideal eS. Define X_R to be the subset of X defined by $X_R = eX$; and let P_L be the partition of X by e. (Note that X_R and P_L depend only on R and L, respectively (Lemma 1.20 and Theorem 1.24).)

1.27. Theorem: Suppose that $S \subset X^X$ has a completely simple minimal ideal $K(S)$. Let R and L be minimal left and right ideals of S, respectively. Then

(i) X_R is invariant under S if and only if $K(S) = R$;

and (ii) P_L is invariant (i.e., P_L, the partition of S by e, say, equals the partition of X by es for every $s \in S$) if and only if $K(S) = L$.

Proof: Fix $e \in E(K(S))$ so that $R = eS$ and $L = Se$.

(i) This is a consequence of Theorem 1.23 (c) and (e), once we note $seX \subset eX$ implies $seX = eX$ (which follows from the fact that eSe is a group).

(ii) This follows from Theorem 1.24 (c) and (e).

We note that the second part of this last result is a variant of the second part of Proposition 5.16 of [Ellis (1969)].

1.28. Theorem: Suppose that $S \subset X^X$ has a completely simple minimal ideal $K(S)$. Also, suppose that there is a non-void invariant subset $Y \subset \cap \{eX \mid e \in E(K(S))\}$. Then

(i) $\cap \{eX \mid e \in E(K(S))\}$ contains a unique maximal invariant subset X_g.

Furthermore,

(ii) X_g is the maximal invariant subset of X such that $S|_{X_g}$ is a group of self-maps of X_g.

Proof: The set X_g which is the union of all the invariant subsets of $\cap \{eX \mid e \in E(K(S))\}$ has all the desired properties. (See the proof of II.1.22, p. 56, of [Berglund and Hofmann (1967)].)

1.29. Notes: (1) If X contains a fixed point z of S, then $\{z\}$ is an invariant subset contained in $\cap \{eX \mid e \in E(K(S))\}$.

(2) On comparing Theorem 1.28 and II.1.22 of [Berglund and Hofmann (1967)], one suspects that the natural context for this type of theorem is in the realm of lattices of subsets of X. In these notes, however, we need only the two cases given.

The next theorem generalizes Theorem 4.10, (ii) - (iv), of [deLeeuw and Glicksberg (1961)]. (To see this, one must note that their semigroup $\overline{S}$ is a compact semitopological semigroup in the weak operator topology, hence has a completely simple minimal ideal (Theorem 2.2 ahead) and that the zero vector is always a fixed point for a semigroup of linear operators on a vector space; of course, deLeeuw and Glicksberg used such facts about $\overline{S}$ in their paper.) See also [Berglund and Hofmann (1967); II.1.23].

1.30. Theorem: Suppose that $S \subset X^X$ has a completely simple minimal ideal $K(S)$. Also, suppose that there is an invariant subset $Y \subset \cap \{eX \mid e \in E(K(S))\}$. Then the following statements are equivalent:

(a) S contains a unique minimal right ideal.

(b) K(S) is a minimal right ideal.

(c) eX is invariant for some $e \in E(K(S))$.

(d) eX is invariant for every $e \in E(K(S))$.

(e) $eX = fX$ for all $e, f \in E(K(S))$.

(f) $eX = X_g$ for every $e \in E(K(S))$.

(g) $eX = X_g$ for some $e \in E(K(S))$.

(h) $S|_{eX}$ is a group for some $e \in E(K(S))$.

(i) $X_g = X_r = \{x \in X \mid y \in Sx \text{ implies } x \in Sy\}$.

Proof: The equivalence of (a) - (d) follows from Theorem 1.27. If (b) holds, then E(K(S)) is a right zero semigroup, which, with Lemma 1.20, implies (e). It follows from (e) that eX is invariant, $e \in E(K(S))$; hence $eX = X_g$ by definition of X_g and (e) implies (f). The implications from (f) to (g) and from (h) to (c) are obvious, while Theorem 1.28 shows (g) implies (h).

To see that (h) implies (i), note first that (h) obviously implies $X_g \subset X_r$ and then that, if $x \in X_r$, there is a $t \in S$ with $x = tex$. But, by Theorem 1.23, $teX = eX$ implies $ete = te$, hence $x = etex$, $X_r = eX$ and Theorem 1.28 finishes the job.

Suppose $X_g = X_r$. Then $X_r \subset eX$ for any $e \in E(K(S))$, while, if $y = tex$, then $ex = (ete)^{-1}y$, i.e., $X_r = eX = X_g$. Thus (i) implies (f).

1.31. <u>Lemma</u>: Suppose that $S \subset X^X$ has a completely simple minimal ideal $K(S)$. Further suppose that $z \in X$ is a fixed point of S. The following assertions concerning a point $x \in X$ are equivalent:

(a) There is some $s \in S$ such that $sx = z$.

(b) $z \in Sx$.

(c) $\{s \in S \mid sx = z\}$ is not empty and is a left ideal of S.

(d) There is a minimal left ideal $L \subset S$ with $Lx = \{z\}$.

(e) There is an idempotent $e \in K(S)$ with $ex = z$.

Proof: Straightforward. (See [Berglund and Hofmann (1967); II.1.24].)

1.32. <u>Definition</u>: Suppose that $S \subset X^X$ has a completely simple minimal ideal $K(S)$, and that $z \in X$ is a fixed point of S. For $e \in E(S)$, define

$$\ker_z e = \{x \in X \mid ex = z\}.$$

1.33. <u>Theorem</u>: Suppose that $S \subset X^X$ has a completely simple minimal ideal $K(S)$. Also suppose that $z \in X$ is a fixed point of S. Then

(i) $\cap \{\ker_z e \mid e \in E(K(S))\}$ contains a unique maximal invariant subspace X_z.

Furthermore,

(ii) X_z is the maximal invariant subset of $\cup \{\ker_z e \mid e \in E(K(S))\}$ such that $S|_{X_z}$ is a semigroup with zero.

Proof: For (i), let X_z be the union of all invariant subsets of $\cap \{\ker_z e \mid e \in E(K(S))\}$. It is clear that each $e \in E(K(S))$ yields a zero for $S|_{X_z}$.

If Y is an invariant subset of

$$\cup \{\ker_z e \mid e \in E(K(S))\}$$

and $S|_Y$ has a zero, then the homomorphism $s \to s|_Y$ must map each $e \in E(K(S))$ onto that zero and, since each $y \in Y$ has $ey = z$ for some $e \in E(K(S))$, the zero of $S|_Y$ is the constant function $y \to z$; and $Y \subset X_z$.

1.34. Theorem: Suppose that $S \subset X^X$ has a completely simple minimal ideal $K(S)$. Also suppose that $z \in X$ is a fixed point of S. Then the following statements are equivalent:

(a) S contains a unique minimal left ideal.

(b) $K(S)$ is a minimal left ideal.

(c) The partition of X by e is invariant for some $e \in E(K(S))$.

(d) The partition of X by e is invariant for every $e \in E(K(S))$.

Moreover, these statements imply the following equivalent statements:

(e) $\ker_z e = \ker_z f$ for all $e, f \in E(K(S))$.

(f) $\ker_z e = X_z$ for every $e \in E(K(S))$.

(g) $\cup \{\ker_z e \mid e \in E(K(S))\}$ is invariant.

And these in turn imply the equivalent statements:

(h) $\ker_z e = X_z$ for some $e \in E(K(S))$.

(i) $S|_{\ker_z e}$ is a semigroup with zero for some $e \in E(K(S))$.

Proof: The equivalence of (a) - (d) follows from Theorem 1.27 (ii). If these hold, then $E(K(S))$ is a left zero semigroup. Thus, if $e, f \in E(K(S))$ and $x \in \ker_z e$, then, since $f = fe$,

$$fx = fex = fz = z$$

and $x \in \ker_z f$. We have established (e).

Suppose (e), and let $e \in E(K(S))$. If $x \in \ker_z e$ and $s \in S$, then, since $(ese)^{-1}s$ is a member f of $E(K(S))$, we have

$$(ese)^{-1}sx = fx = z;$$

hence $esx = (ese)(ese)^{-1}sx = (ese)fx = z$, $\ker_z e$ is invariant and we have (f).

It is obvious that (f) implies (e) and (h), and Theorem 1.33 shows that (f) is equivalent to (g) and (h) is equivalent to (i).

1.35. Comment: Obviously, one would like to conclude that all the statements (a) - (i) are equivalent, as they are if the elements of S are linear operators on a vector space X and z is the zero vector [Berglund and Hofmann (1967); II.1.26]. That this is not possible follows from the example of a (non-trivial) right zero semigroup S acting on $X = S \cup \{u\}$, where u is an adjoined identity.

1.36. Lemma: Suppose that $S \subset X^X$ has a completely simple minimal ideal $K(S)$ which contains no nontrivial groups, i.e., $K(S) = E(K(S))$. Also, suppose that there is a non-void invariant subset $Y \subset \cap \{eX \mid e \in E(K(S))\}$. Then the following statements about a point $x \in X$ are equivalent:

(a) $x \in X_g$.

(b) x is a fixed point of S.

Proof: By construction, X_g is the union of all the invariant subsets of $\cap \{eX \mid e \in E(K(S))\}$. Hence, if $x \in X_g$, then $ex = x$ for every $e \in E(K(S))$. Therefore, if $s \in S$, then $sx = sex$; but se is in $E(K(S)) = K(S)$ and we have that

$$sx = sex = x$$

for every $s \in S$. That (b) implies (a) is trivial.

1.37. <u>Theorem</u>: Suppose $S \subset X^X$ has a completely simple minimal ideal $K(S)$ which contains no non-trivial groups.

1. Suppose X contains at least one fixed point for S. Then the equivalent statements of Theorem 1.30 are equivalent to

   (j) Sx contains a fixed point of S for every $x \in X$.

2. S has a zero element if and only if Sx contains a unique fixed point of S for every $x \in X$.

Proof: If S contains a unique minimal right ideal, then $E(K(S))$ is a right zero semigroup. Let $e \in E(K(S))$. Since $eX = X_g$ and X_g is invariant, we have that, for any $x \in X$ and $s \in S$,

$$sex = esex = ex,$$

since the group $eSe = \{e\}$. Thus, ex is a fixed point contained in Sx; and (a) and (f) of Theorem 1.30 imply (j) above.

To see that (j) implies (i) of Theorem 1.30 we note first that the equality $K(S) = E(K(S))$ and Lemma 1.36 imply that $X_g = X_p$, the set of fixed points of S. Hence, $X_p \subset X_r$. Suppose $x \in X_r$. Sx contains a fixed point tx. Since $x \in X_r$, there is an $s \in S$ with $x = stx$; but since tx is a fixed point point, $stx = tx$. Thus, x is a fixed point in Sx, and therefore $x \in X_g$ as desired.

Now, if S has a zero e, then clearly ex is a fixed point for every $x \in X$. Moreover, if tx is a fixed point of S, then $tx = etx = ex$, so the fixed point in Sx is unique.

If, on the other hand, Sx contains a unique fixed point for each $x \in X$, it follows that $ex = fx$ for all $x \in X$ and for all $e, f \in E(K(S)) = K(S)$; hence K(S) has only one element.

1.38. <u>Theorem</u>: Suppose $S \subset X^X$ has a completely simple minimal ideal K(S). Then statements (a) and (b) are equivalent, as are statements (a') and (b').

(a) S contains a right zero e.

(b) Sx contains a fixed point of S for every $x \in X$.

(a') S contains a zero e.

(b') Sx contains a unique fixed point of S for every $x \in X$.

Proof: If e is a right zero, then ex is a fixed point of S for any $x \in X$. Thus, (a) implies (b) and (a') implies (b') follow from 1.37.

(b) implies (a). Let $X_p = \{x \in X \mid x \text{ is a fixed point of } S\}$. Then $X_p \subset X_g \subset X_r$ and the containment $X_r \subset X_p$

follows as in 1.37, whence, by 1.30, K(S) is a minimal right ideal and, if $e \in E(K(S))$, $K(S) = eS \times E(Se)$ by 1.13. Note that, since $X_g = X_p$, we also have $ex \in X_p$ for every $x \in X$ by 1.30 (f). And, if $t_1, t_2 \in Se$, then

$$t_1x = t_1ex = ex = t_2ex = t_2x$$

for every $x \in X$; thus $t_1 = t_2$ and Se is trivial.

That (b') implies (a') now follows readily.

2. COMPACT RIGHT TOPOLOGICAL SEMIGROUPS

Let S be a compact (Hausdorff) right topological semigroup, i.e., ρ_t: $s \to st$ is a continuous map of S into S for all $t \in S$. The set of idempotents of S is denoted by E(S) and we let $\Lambda = \Lambda(S)$ be the subset $\{s \in S \mid \lambda_s: S \to S$ is continuous$\}$. The following result leads to the complete algebraic description of the minimal ideal $K = K(S)$ of S. Ruppert (1973) observed this, and many of the results recorded here appear in his paper; e.g., 2.2, 2.7 - 2.10, 2.12 and 2.13 ahead.

2.1. Proposition [Ellis (1969); Corollary 2.10]: Let S be a compact right topological semigroup. Then there exists an idempotent in S.

Proof: Consider the collection I of all closed non-void subsets J of S satisfying $J^2 \subset J$. I is non-void since $S \in I$. If I is ordered downward by inclusion, then, by the compactness of S, I contains a minimal member H. Let $s \in H$. Then $Hs = \{\rho_s t \mid t \in H\}$ is non-void and compact, and $HsHs \subset HHHs \subset Hs$; hence, by the minimality of H, $Hs = H$. Therefore, there exists $t \in H$ with $ts = s$. Since $t \in W = \rho_s^{-1}(s) \cap H = \{r \mid rs = s\} \cap H$, W is non-void, closed and satisfies $W^2 \subset W$. Hence $W = H$, $s \in W$ and $s^2 = s$.

2.2. Theorem: Let S be a compact right topological semigroup.

(i) Every left ideal of S contains a minimal left ideal. The minimal left ideals of S are closed.

(ii) S has a smallest two-sided ideal $K = K(S)$.

(iii) K contains idempotents and, for an idempotent $e \in S$, the following are equivalent:

(a) $e \in K$.

(b) $K = SeS$.

(c) Se is a minimal left ideal.

(d) eS is a minimal right ideal.

(e) eSe is a maximal subgroup of S.

(iv) Every minimal left ideal is of the form Se for some idempotent $e \in K$; every minimal right ideal is of the form eS for some idempotent $e \in K$.

(v) $K = \cup\{eSe \mid e \in E(K)\} = \cup\{eS \mid e \in E(K)\}$
$= \cup\{Se \mid e \in E(K)\}$.

(vi) All maximal subgroups of K are algebraically isomorphic; maximal subgroups in the same minimal right ideal are topologically isomorphic.

(vii) For each idempotent $e \in K$, K is algebraically isomorphic to $E(Se) \times eSe \times E(eS)$ with multiplication $(u,v,w)(x,y,z) = (u,vwxy,z)$. And minimal right ideals are algebraically isomorphic to the right-group $eSe \times E(eS)$, minimal left ideals to the left-group $E(Se) \times eSe$.

Proof: (i) follows from the compactness of S and the fact that ρ_t is continuous for each $t \in S$. Proposition 2.1 then implies that each minimal left ideal contains an idempotent. The rest follows from Theorem 1.16, except the second statement of (vi), which then follows from the continuity of the maps ρ_t, $t \in S$.

Part of the next result concerns analogues for right ideals of the previous result. (iii) can be thought of as a generalization of (2.10) Theorem of [Witz (1964)].

2.3. Theorem: Let S be a compact right topological semigroup.

(i) Every right ideal of S contains a minimal right ideal.

(ii) Every closed right ideal contains a minimal closed right ideal.

(iii) In the event that all the maximal subgroups of K are closed, there is a 1-1 correspondence between the minimal closed right ideals of S and the minimal right ideals of S given by
$R \to R \cap K \to \overline{R \cap K} = R$.

Proof: (i) and (ii) are obvious. Suppose all the maximal subgroups of K are closed and R is a minimal closed right ideal. Clearly $R \cap K$ is a right ideal and $\overline{R \cap K} = R$. We prove $R \cap K$ is minimal by showing that, for each minimal left ideal L of K, $R \cap K \cap L = R \cap L$ contains exactly one maximal subgroup of K. For, suppose $R \cap L$ contains two such subgroups G_1 and G_2 with identities e_1 and e_2. Then $\overline{e_1 S} = R$ and there is a net $\{t_\alpha\} \subset S$ with $e_1 t_\alpha \to e_2$, which implies $e_1 t_\alpha e_1 \to e_2 e_1 = e_2$, this last equality by Theorem 2.2, (vii). But $\{e_1 t_\alpha e_1\} \subset G_1$ (Theorem 2.2, (iii)), which is closed; so $e_2 \in G_1$, which is a contradiction.

Similarly, if R is a minimal right ideal, $\overline{R} \cap L$ cannot contain two maximal subgroups of K for any minimal left ideal L; so $\overline{\overline{R} \cap K} = R$.

2.4. Remarks: (a) Minimal right ideals need not be closed (Examples V.1, 1, 3 and 10).

(b) Maximal subgroups of $K = K(S)$ need not be closed or pairwise topologically isomorphic (Examples V.1, 2, 3 and 10). In Example V.1.10, Λ is dense in S as well.

(c) If R(L) is a minimal right (left) ideal, so is sR (Ls) for any $s \in S$ (Lemma 1.12). If R is closed as well, sR need not be (Example V.1.3), but is if $s \in \Lambda$.

(d) The proof of Theorem 2.3 (iii) uses the fact that the closure of a right ideal is also a right ideal. Is the closure of a left ideal also a left ideal? See Example V.1.11.

We next show that, when Λ is a dense commutative subset of S (i.e., when S has dense center), the existence of maximal subgroups of K that are not closed or not pairwise topologically isomorphic is related to other properties of K. Proposition 2.6 is essentially due to Butcher (1975).

2.5. Proposition: Let S be a compact right topological semigroup such that Λ is a dense commutative subset of S and suppose S has a minimal right ideal R that is closed. Then $R = K$ and all the maximal subgroups of K are closed and pairwise topologically isomorphic.

Proof: Since $sR = Rs \subset R$ for all $s \in \Lambda$, it follows that $SR \subset \overline{R} = R$. Thus $R = K$ and each maximal subgroup of K is a minimal left ideal. The rest follows from Theorem 2.2, (i) and (vi).

2.6. Proposition: Let S be a compact right topological semigroup such that Λ is a dense commutative subset of S.

Then each maximal subgroup of K is dense in the minimal left ideal containing it. Hence, K is a minimal right ideal if and only if K has a maximal subgroup that is closed; in this case, all maximal subgroups of K are closed and pairwise topologically isomorphic.

Proof: Let G be a maximal subgroup of K contained in a minimal left ideal L of S. Then st = ts for all $s \in \Lambda$, $t \in S$, which implies $s\overline{G} = \overline{G}s \subset \overline{G}$ for all $s \in \Lambda$; hence $S\overline{G} \subset \overline{G}$, which must then be equal to L. The remaining assertions follow readily.

2.7. <u>Proposition</u>: Let S be a compact right topological semigroup, $\mathcal{R}$ a closed congruence in S. Then $S/\mathcal{R}$ (with the quotient topology) is again a compact right topological semigroup.

Proof: Since $\mathcal{R}$ is a congruence, $S/\mathcal{R}$ is a semigroup with multiplication $[s]_{\mathcal{R}}[t]_{\mathcal{R}} = [st]_{\mathcal{R}}$ and $S/\mathcal{R}$ is compact, since $\mathcal{R}$ is closed. The continuity, for each $t \in S$, of $\rho_{[t]_{\mathcal{R}}}: S/\mathcal{R} \to S/\mathcal{R}$ follows from the fact that the map $s \to [st]_{\mathcal{R}}$ is continuous.

2.8. <u>Proposition</u>: Let S be a compact right topological semigroup. Then $\Lambda = \{s \in S \mid \lambda_s: S \to S \text{ is continuous}\}$ is void or a subsemigroup of S. If S is a group, Λ is a subgroup of S.

Proof: If Λ is not void, it is a subsemigroup of S, since the composition of continuous maps is continuous. If S is a group, the identity e is in Λ and the compactness of S implies $\lambda_{s^{-1}}$ is continuous whenever λ_s is.

2.9. Theorem: Let S be a left simple compact right topological semigroup with $\Lambda(S) = \{s \in S \mid \lambda_s: S \to S \text{ is continuous}\} \neq \emptyset$. Then the following assertions hold.

(i) E(S) is a compact topological left zero semigroup.

(ii) A maximal subgroup G of S is closed if and only if $G \cap \Lambda(S) \neq \emptyset$. In this case $G \cap \Lambda(S)$ is a group and equals $\Lambda(G) = \{s \in G \mid \lambda_s: G \to G \text{ is continuous}\}$.

(iii) Any two closed maximal subgroups G_1 and G_2 of S are topologically isomorphic as are $\Lambda(G_1)$ and $\Lambda(G_2)$.

(iv) The following assertions are equivalent.

(a) For any $e \in E(S)$, S is topologically isomorphic to the direct product (left-group) $E(S) \times eS$.

(b) The map $\psi: E(S) \times S \to S$, $\psi(f,s) = fs$ is continuous.

Proof: Suppose $s \in \Lambda(S)$ and $\{e_\alpha\}$ is a net in E(S), $e_\alpha \to f \in S$. There are maximal subgroups G_1 and G_2 of S with identities e_1 and e_2, respectively, such that $s \in G_1$, $f \in G_2$. Then $s = se_\alpha$ for all α and $se_\alpha \to sf$, hence $s = sf$; and $f = e_2f = e_2e_1f = e_2$, since $e_1f = s^{-1}sf = e_1$ (where s^{-1} is the inverse in G_1 of s). This proves (i).

If G is a maximal subgroup of S and $s \in G \cap \Lambda(S)$, then $G = sS$ and is compact. Since $\Lambda(S) \neq \emptyset$, $G \cap \Lambda(S) \neq \emptyset$ for at least one maximal subgroup G of S, and $\Lambda(G)$ contains $G \cap \Lambda(S)$, hence is a group (Proposition 2.8) and the identity e of G is in $\Lambda(G)$. But any $s \in \Lambda(G)$ is also in $\Lambda(S)$. For, if $t \in G \cap \Lambda(S)$, then $\lambda_s\colon S \to S$ maps S into G and is a composition of continuous maps, $\lambda_s = \lambda_s \circ \lambda_{t^{-1}} \circ \lambda_t$, λ_t mapping S into G. And, if G_1 is any other maximal subgroup of S, λ_e is a continuous isomorphism of G_1 onto G which has a continuous inverse if (and only if) G_1 is compact. It is clear that λ_e injects $\Lambda(G_1)$ into $\Lambda(G)$ and, by symmetry, λ_e effects a topological isomorphism between $\Lambda(G_1)$ and $\Lambda(G)$ (when G_1 is compact). This completes the proof of (ii) and (iii).

If (b) of (iv) is satisfied, then $E(S) \subset \Lambda(S)$, all the maximal subgroups of S are closed and, if $G = eS$ is one of those subgroups, then the algebraic isomorphism (Theorem 2.2, (vii)) of $E(S) \times eS$, which is compact, onto S, $(f,es) \to fes = fs$, is continuous, hence a topological isomorphism. That (a) implies (b) is easier.

Examples V.1, 2 and 3, are relevant to Theorem 2.9; e.g., 2 has $\Lambda = \emptyset$ and has two closed maximal subgroups that are not topologically isomorphic, while in 3 the closed maximal subgroup equals Λ and the one that is not closed equals $S\backslash\Lambda$. See also Example V.1.10. The next theorem shows the situation for a right simple compact right topological semigroup is less complicated.

2.10. <u>Theorem</u>: Let S be a right simple compact right topological semigroup. Then

(i) All maximal subgroups of S are closed.

(ii) E(S) is a compact topological right zero semigroup.

(iii) The following assertions are equivalent.

(a) For any $e \in E(S)$, S is topologically isomorphic to the direct product (right-group) $Se \times E(S)$.

(b) The map $\psi: S \times E(S) \to S$, $\psi(s,f) = sf$ is continuous.

Proof: Since the maximal subgroups are the minimal left ideals, (i) holds. If a net $\{e_\alpha\} \subset E(S)$ converges to a member f of a maximal subgroup G of S with identity e, then $e_\alpha f = e_\alpha ef = ef = f$ for all α and $e_\alpha f \to f^2$, which implies $f^2 = f$, i.e., $f = e$. This shows $E(S)$ is closed and proves (ii). (iii) is proved like the analogous part of Theorem 2.9.

2.11. Remark: When S is a right simple compact semitopological semigroup, the maximal subgroups of S (which are the minimal left ideals of S) are compact topological groups (by Ellis' theorem). If $e \in E(S)$, Se is one such group and the algebraic isomorphism (Theorem 2.2 (vii)) of $Se \times E(S)$ and S, $(s,f) \to sf$, which is separately continuous, is jointly continuous (again, by Ellis' theorem), hence a homeomorphism. Thus S is a topological semigroup. See [Berglund and Hofmann (1967); II.2.3].

2.12. Theorem: For a compact right topological semigroup S the following assertions are equivalent.

(i) The maps $\{\rho_s \mid s \in S\}$ are equicontinuous.

(ii) There are a compact topological semigroup Γ, an algebraic antiisomorphism γ of S onto a dense subset of Γ and a continous map $\delta\colon \Gamma \to S$ with $\delta(\gamma(s)) = s$ for all $s \in S$.

Proof: In order that the map $s \to \rho_s$ will be injective we need the semigroups we deal with to have at least left identities. Accordingly, for a compact right topological semigroup T, let $T^1 = T$ if T has a left identity 1; otherwise, let $T^1 = T \cup \{1\}$, i.e., adjoin an identity 1 as a discrete point, $t1 = 1t = t$ for all $t \in T^1$. In particular, T^1 is also a compact right topological semigroup.

Suppose now that (i) is satisfied. Then $\{\rho_s\colon S^1 \to S^1 \mid s \in S^1\}$ is equicontinuous and, by I.1.5, has compact closure Ω in $C(S^1,S^1)$, the semigroup of all continuous maps of S^1 into S^1, $C(S^1,S^1)$ having the topology of uniform convergence on S^1, which makes it a topological semigroup. Also, the map $\gamma\colon s \to \rho_s$ is an antiisomorphism of S onto $\gamma(S) \subset \Omega$ whose closure Γ is a compact topological semigroup. The map δ is defined for $f \in \Gamma$ by $\delta(f) = f(1)$. Clearly δ is continuous and $\delta(\gamma(s)) = s$ for all $s \in S$; hence, δ maps $\Gamma(S)$ onto S.

On the other hand, if (ii) is satisfied, we note first that $\{\rho_f\colon \Gamma \to \Gamma \mid f \in \Gamma\}$ is (uniformly) equicontinuous, since Γ is a compact topological semigroup. Thus, given any entourage U ($\subset \Gamma \times \Gamma$ of the unique uniformity $\mathcal{U}$) of Γ, we can find another entourage U' such that $(gf,hf) = (\rho_f g,\rho_f h) \in U$ whenever $f \in \Gamma$, $(g,h) \in U'$. Also, since δ is a continuous, hence uniformly continuous, map of Γ into S, given any entourage U of S, we can find an entourage U' of Γ such that $(\delta(g'),\delta(h')) \in U$ whenever $(g',h') \in U'$. Combining these last two sentences and replacing f, g, h, with

$\gamma(s)$, $\gamma(t)$, $\gamma(u)$ we have that, given entourage U of S, we can find an entourage U' of Γ such that $(\rho_t s, \rho_u s) = (st, su) \in U$ for all $s \in S$ whenever $(\gamma(t), \gamma(u)) \in U'$, i.e., the map, $\gamma(S) \to C(S,S) : \gamma(t) \to \rho_t$, is uniformly continuous, $C(S,S)$ having the uniformity of uniform convergence on S. Thus $\{\rho_t \mid t \in S\}$ is the uniformly continuous image of the totally bounded set $\gamma(S) \subset \Gamma$, and hence is totally bounded and equicontinuous.

2.13. Theorem: Let S_1 and S_2 be compact right topological semigroups that satisfy one (hence both) of the conditions of Theorem 2.12 and have associated compact topological semigroups Γ_1 and Γ_2 and antiisomorphisms $\gamma_1 : S_1 \to \Gamma_1$ and $\gamma_2 : S_2 \to \Gamma_2$, respectively. If ν is a continuous homomorphism of S_1 onto S_2, there is a continuous homomorphism ω of Γ_1 onto Γ_2 such that $\omega(\gamma_1(s)) = \gamma_2(\nu(s))$ for all $s \in S_1$.

Proof: (We regard Γ_i as a compact subsemigroup of $C(S_i^1, S_i^1)$ furnished with the topology of uniform convergence on S_i^1, $i = 1, 2$.) Since ν is continuous and hence uniformly continuous, given entourage U of S_2, we can find an entourage U' of S_1 such that $(\rho_{\nu(t)}\nu(s), \rho_{\nu(u)}\nu(s)) = (\nu(st), \nu(su)) \in U$ for all $s \in S_1^1$ whenever $(\rho_t s, \rho_u s) = (st, su) \in U'$ for all $s \in S_1^1$. Since ν maps S_1 onto S_2, this says that the homomorphism $\gamma_1(S_1) \to \gamma_2(S_2) : \gamma_1(s) \to \gamma_2(s)$ is uniformly continuous; its extension to $\Gamma_1 = \overline{\gamma_1(S_1)}$ is the required homomorphism.

We conclude this section with a result relating back to Theorem 1.38.

2.14. Theorem: Let (X,S) be a flow with enveloping semigroup E (I.2.1).

1. The minimal ideal $K(E)$ is a right zero semigroup if and only if $\overline{Sx}$ contains a fixed point for every $x \in X$.
2. E contains a zero element if and only if $\overline{Sx}$ contains a unique fixed point for every $x \in X$.

Proof: Since $\overline{Sx} = Ex$, these results follow from Theorem 1.38.

3. COMPACT RIGHT TOPOLOGICAL GROUPS

It follows from a result of R. Ellis that a (locally) compact semitopological semigroup that is a group is in fact a topological group. (See [Namioka (1974); §3].) However, there do exist compact non-topological right topological groups; see Examples V.1, 5 and 7 (7 in conjunction with Theorem 3.8 ahead).

3.1. Theorem [Ruppert (1973), Namioka (1974)]: If G is a metric compact right topological group, then $\Lambda = \{s \in G \mid \lambda_s: G \to G$ is continuous$\}$ is closed and a topological group. In particular, if $\overline{\Lambda} = G$, then G is a topological group. (The conclusions here still hold if G is a metric locally compact right topological group.)

Proof: If $\{s_n\} \subset \Lambda$ and $s_n \to s \in G$, then $\{\lambda_{s_n}\}$ converges to λ_s pointwise on G, and there must be at least one point in G where λ_s is continuous [Dugundji (1966); p. 277]; since G is a group, λ_s is continuous at every point of G, hence $s \in \Lambda$.

We state the following theorem without proof (noting that its conclusion is trivial for the compact right topological group of Example V.1.5, where the measure whose existence is asserted can be just Haar measure on the closed subgroup $\Lambda(G \times F') = G \times F''$, where $F'' = \{\psi' \in F' \mid \psi'$ is continuous$\}$.

3.2. Theorem [Namioka (1972)]: Let G be a compact right topological group. Then G admits a probability measure μ

with the invariance property, $\mu(f) = \mu(L_s f)$ for all $f \in C(G)$, $s \in \Lambda(G)$.

3.3. <u>Theorem</u> [Ruppert (1975)]: The following assertions about a compact right topological group G are equivalent.

(i) The maps $\{\rho_s \mid s \in G\}$ are equicontinuous.

(ii) The map $(s,t) \to st$ is continuous at (e,e).

(iii) The sets $\{\{(s,t) \mid st^{-1} \in V\} \subset G \times G \mid V$ is a neighbourhood of $e \in G\}$ form a base for the uniformity of G.

(iv) There are a compact topological group Γ and an algebraic antiisomorphism γ of G onto a dense subgroup M of Γ. Also, there is a continuous map $\delta: \Gamma \to G$ with $\delta(\gamma(s)) = s$ for all $s \in G$; the kernel $H = \{f \in \Gamma \mid \delta(f) = e\}$ is a compact subgroup of Γ and δ induces a homeomorphism between the quotient space Γ/H and G.

When these hold, the map $s \to s^{-1}$ is continuous at e.

Proof: <u>(i) implies (ii)</u>. Let $\{s_\alpha\}$ and $\{t_\beta\}$ be nets in G converging to the identity $e \in G$ and let U_1 be a member of the uniformity $\mathcal{U}$ of G. Let $U_2 \in \mathcal{U}$ satisfy $U_2^2 \subset U_1$. By hypothesis, there exist α_0 and β_0 such that $(t_\beta, e) \in U_2$, $(s_\alpha s, s) \in U_2$ for all $\alpha \geq \alpha_0$, $\beta \geq \beta_0$ and for all $s \in S$. In particular $(s_\alpha t_\beta, t_\beta) \in U_2$, hence $(s_\alpha t_\beta, e) \in U_2^2 \subset U_1$, for all $\alpha \geq \alpha_0$, $\beta \geq \beta_0$.

<u>(ii) implies (iii)</u>. From (ii) it follows that every net $\{s_\alpha\}$ converging to e has a subnet $\{s_{\alpha_\beta}\} = \{s_\beta\}$ such that $s_\beta^{-1} \to e$. Hence inversion in G, $s \to s^{-1}$, is continuous at e. (The last statement of the theorem is thus dealt with.) Now let V be a neighbourhood of e. Since

$(s,t) \to st$ is continuous at (e,e), there is a neighbourhood W of e with $W^2 \subset V$. Since inversion is continuous at e we may assume $W = W^{-1}$. It follows that $st^{-1} \in V$ whenever $s, t \in Wt_0$ for some $t_0 \in G$. Thus $N_V = \{(s,t) \mid st^{-1} \in V\} \supset \cup\{(Wt_0 \times Wt_0) \mid t_0 \in G\}$, which is a neighbourhood of the diagonal D of $G \times G$, and is a member of $\mathcal{U}$. Since $\cap\{N_V \mid V \text{ is a neighbourhood of } e\} = D$, $\{N_V \mid V$ is a neighbourhood of $e\}$ is a base for $\mathcal{U}$.

<u>(iii) implies (i)</u>. If $(t,u) \in N_V$, then $(\rho_s t, \rho_s u) = (ts,us) \in N_V$ for all $s \in G$.

<u>(i), (ii) and (iii) imply (iv)</u>. After citing Theorem 2.12, we note that Γ will be a group since it is compact and the closure in a topological semigroup of a group. Also $H = \{f \in \Gamma \mid \delta(f) = e\}$ is a compact group, for, if $f, g \in H$, there are nets $\{\rho_{s_\alpha}\}$ and $\{\rho_{t_\beta}\}$ in M converging to f and g, respectively; hence $\{s_\alpha\}$, $\{t_\beta\}$, $\{s_\alpha^{-1}\}$ and $\{t_\beta s_\alpha^{-1}\}$ all converge to $e \in G$ and $t_\beta s_\alpha^{-1} = \delta(\rho_{s_\alpha}^{-1}\rho_{t_\beta}) \to \delta(f^{-1}g) = e$. Clearly $M \cap H = \{e\}$. To see $\Gamma = MH$, let $f \in \Gamma$; $\rho_{s_\alpha} \to f$ for a net $\{s_\alpha\} \subset G$. Then, for some $s \in G$, $s_\alpha \to s$, and $f = \rho_s \rho_s^{-1} f$, where

$$\delta(\rho_s^{-1} f) = \lim_\alpha \delta(\rho_s^{-1}\rho_{s_\alpha}) = \lim_\alpha s_\alpha s^{-1} = e,$$

i.e., $\rho_s^{-1} f \in H$. It is clear that δ factors through Γ/H; it follows from the openness of the quotient map, $\Gamma \to \Gamma/H$, that the induced bijection $\Gamma/H \to G$ is continuous, hence a homeomorphism.

<u>(iv) implies (i)</u>. This follows from Theorem 2.12.

3.4. Remarks: 1. Another way of getting the group Γ from the group G is to complete G in the uniformity $\mathcal{U}_1$, which has as a base $\{(s,t) \mid st^{-1} \in rVr^{-1}, r \in A\}$, where V is a neighbourhood of $e \in G$ and $A \subset G$ is finite. $\mathcal{U}_1$ is a stronger uniformity than the original uniformity of G and is the uniformity of pointwise convergence on G for the maps $\{\rho_s \mid s \in G\}$.

2. $\Lambda(G) = \{s \in G \mid \lambda_s: G \to G$ is continuous$\}$ is the image under δ of the normaliser N(H) of H in Γ; therefore $\Lambda(G)$ is a compact topological group. For, let $f \in \Gamma$ with $\delta(f) \in \Lambda(G)$ and let $h \in H$. Then there is a net $\{m_\alpha\} \subset M$ converging to h and $\{\delta(m_\alpha f)\} = \{\delta(f)\delta(m_\alpha)\}$ converges to $\delta(hf)$; and $\delta(m_\alpha f) = \delta(f)\delta(m_\alpha) \to \delta(hf) = \delta(f)$ since $\delta(f) \in \Lambda(G)$ and $\delta(m_\alpha) \to \delta(h) = e$, i.e., $hf \in fH$, $f \in N(H)$.

On the other hand, let $f \in N(H)$. If $\{s_\alpha\} \subset G$ is a net converging to $s \in G$, $\{\rho_{s_\alpha}\}$ has a subnet $\{\rho_{s_\beta}\}$ converging to $mh \in MH = \Gamma$. Therefore $\delta(\rho_{s_\beta} f) = \delta(f)s_\beta \to \delta(mhf) = \delta(fh)\delta(m) = \delta(f)s$, i.e., $\delta(f) \in \Lambda(G)$.

3. $\{e\}$ is the only normal subgroup of Γ contained in H.

4. The map $\delta: \Gamma \to G$ is a homomorphism if and only if G is a topological group, in which case $G \cong \Gamma$.

We quote without proof the following theorem.

3.5. Theorem [Ruppert (1974, 1975)]: A right topological group defined on the topological space R is a topological group isomorphic to the group of real numbers. The analogous conclusion with R replaced by R/Z holds. A compact right topological group G with $\{\rho_s \mid s \in G\}$ equicontinuous, which is defined on a compact connected manifold of dimension 1, 2 or 3 is a topological group.

Compact right topological semigroups and compact right topological groups arise naturally in the study of flows. Flows have been discussed in §I.2 (and the reader is referred there for definitions and basic results); for our purposes here we consider a flow to be a pair (X,S), where X is a compact Hausdorff space and S is a semigroup of continuous transformations of X. (Thus, we are suppressing notation for the function from $S \times X$ into X.) The following elegant results appear in [Ellis (1969)], where S is assumed to be a group; the proofs there work for semigroups.

3.6. Theorem: A flow (X,S) is distal if and only if its enveloping semigroup $\mathcal{E}$ is a group. If (X,S) is distal, then $\mathcal{E}$ consists of continuous transformations of X if and only if S is equicontinuous; when this is the case $\mathcal{E}$ is a topological group.

Proof: Let (X,S) be distal and let e be an idempotent in the minimal ideal $K(\mathcal{E})$ of $\mathcal{E}$; let $x \in X$ and $y = ex$. If $\{s_\alpha\}$ converges in $\mathcal{E}$ to e, then

$$ey = \lim_\alpha s_\alpha y = \lim_\alpha s_\alpha ex = eex = ex = \lim_\alpha s_\alpha x,$$

i.e., $x = y$, e is the identity in X^X and $\mathcal{E}$ is a group. On the other hand, if $\mathcal{E}$ is a group and $\lim_\alpha s_\alpha x = \lim_\alpha s_\alpha y = z$ for some $x, y, z \in X$, $\{s_\alpha\} \subset S$, then there is an $f \in \mathcal{E}S$ with $fx = fy = z$ and $x = f^{-1}fx = f^{-1}fy = y$, i.e., (X,S) is distal.

Suppose (X,S) is distal and $\mathcal{E}$ is a group. It is clear that $\mathcal{E}$ consists of continuous maps (and is a topological group) if S is equicontinuous: $\mathcal{E}$ is a topological semigroup by I.2.2 and it is easily verified that a compact topological semigroup with a dense subgroup must itself be a topological group.

Conversely, if $\mathcal{E}$ consists of continuous maps, then $\mathcal{E}$ is a semitopological semigroup (I.2.2) and the map $\mathcal{E} \times X \to X$, $(f,x) \to fx$ is separately continuous. Ellis' theorem (see [Namioka (1974)]) then implies this map is jointly continuous, from which it follows that $\mathcal{E}$, and hence S, is equicontinuous.

In the second half of the last theorem where $\mathcal{E}$ is a group and S is equicontinuous, then $\mathcal{E}$ is a compact topological group in the topology of uniform convergence on X. If S is not equicontinuous, then $\mathcal{E}$ is not compact in the topology of uniform convergence on X, but may still be a topological group with respect to the topology of pointwise convergence on X, in which it is compact. The next theorem tells when $\mathcal{E}$ is a topological group in this topology, and is hinted at in [Namioka (1972)]; for it we need a definition.

3.7. <u>Definition</u>: Let (X,S) be a flow. A subset Y of X is called <u>minimal</u> provided it is closed, $SY \subset Y$ and Sy is dense in Y for every $y \in Y$.

It follows by compactness that, for every flow (X,S), X will contain minimal subsets; and, if (X,S) is distal, then X decomposes into a disjoint union of minimal subsets,

$$X = \cup X_\gamma = \cup \overline{Sx_\gamma} = \cup \mathcal{E}x_\gamma,$$

where $x_\gamma \in X_\gamma$ for each γ. Also, if Y is a minimal subset of X, then the restriction of S to Y determines a flow (Y,S_Y), where S_Y is a homomorphic image of S. This flow will be distal if (X,S) is distal.

3.8. Theorem: Let (X,S) be a distal flow with enveloping semigroup $\mathcal{E}$. Suppose S is not equicontinuous. Then $\mathcal{E}$ is not a compact topological group in the topology of uniform convergence on X; but, furnished with the topology of pointwise convergence on X, it is a compact topological group if and only if the restriction of S to each minimal subset Y of X determines a flow (Y,S_Y) with S_Y equicontinuous.

Proof: It has been noted that $\mathcal{E}$ is not compact in the topology of uniform convergence on X if S is not equicontinuous. Suppose $X = \cup X_\gamma$ is a decomposition of X into disjoint minimal subsets and suppose $S_\gamma = S_{X_\gamma}$ is equicontinuous for each γ; then the enveloping semigroup $\mathcal{E}_\gamma$ of (X_γ, S_γ) is a compact topological group (in the topology of pointwise convergence on X_γ or the topology of uniform convergence on X_γ) by Theorem 3.6. Hence the canonical image of S in X^X is contained in the product of compact topological groups,

$$\pi \mathcal{E}_\gamma (\subset \pi X_\gamma^{X_\gamma} \subset X^X),$$

and $\mathcal{E}$, its closure there, is a compact topological group.

Conversely, suppose Y is a minimal subset of X and S_Y is not equicontinuous. Then there are an $x \in Y$, a member $\mathcal{W}$ of the (unique) uniformity of Y and nets $\{x_\beta\} \subset Y$, $\{s_\beta\} \subset S_Y$ with $x_\beta \to x$ and

$$(s_\beta x_\beta, s_\beta x) \notin \mathcal{W}$$

for every β. Let E_Y be the enveloping semigroup of (Y,S_Y). We may assume $s_\beta \to \psi \in E_Y$ and, since $E_Y x = Y$, we can find a net $\{\psi_\beta\} \subset E_Y$ such that $\psi_\beta x = x_\beta$ for all β; we may also assume $\psi_\beta \to \psi_0 \in E_Y$. Then

$$s_\beta x \to \psi x = \psi\psi_0 x \text{ and } s_\beta x_\beta = s_\beta \psi_\beta x \not\to \psi\psi_0 x.$$

Thus E_Y is not a topological group. Since the map restricting members of E to Y is a continuous homomorphism of E onto E_Y, E is also not a topological group.

3.9. Remarks: Examples V.1, 6 and 7, are of distal flows that are not equicontinuous. They illustrate the final conclusion of the previous theorem.

4. COMPACT AFFINE RIGHT TOPOLOGICAL SEMIGROUPS

In this section we study affine semigroups. We recall from Definition I.1.2 that an affine semigroup is a semigroup T that is also a convex subset of a (locally convex Hausdorff topological) vector space such that the maps $\rho_t: s \to st$ and $\lambda_t: s \to ts$ from T into T are affine for every $t \in T$ (e.g.,

$$\rho_t(as_1 + (1 - a)s_2) = a\rho_t(s_1) + (1 - a)\rho_t(s_2),$$

$s_1, s_2, t \in T$, $a \in [0,1]$).

The first result we give here asserts that every finite dimensional affine semigroup is topological. In conjunction with this, as well as noting Example V.1.4, we recall the example on p. 156 of Berglund and Hofmann (1967) where the (compact convex) unit cube $[0,1]^3$ is given a multiplication that makes it a semitopological non-topological (and hence non-affine) semigroup; this is also (primarily, one could say) an example of a compact semitopological semigroup whose minimal two-sided ideal is not closed.

4.1. Theorem: Every finite dimensional affine semigroup is topological.

Proof: Let T be an n-dimensional affine semigroup. Let

$$\{a_1, a_2, \ldots, a_n\} \subset T$$

be a basis for the linear span of T. Since all vector topologies for finite dimensional spaces are equivalent, we can use the metric

$$d(s,t) = \max\{|\sigma_i - \tau_i| \mid 1 \le i \le n\}$$

for s and t in the linear span of T,

$$s = \sum_1^n \sigma_i a_i, \quad t = \sum_1^n \tau_i a_i .$$

We must show that, if s_0, t_0, s, t $\in$ T and $d(s,s_0) \to 0$, $d(t,t_0) \to 0$, then $d(st,s_0t_0) \to 0$. This is achieved by noting that, if

$$a_i a_j = \sum_1^n \rho_k^{ij} a_k, \quad 0 \le i,\ j \le n,$$

with $M = \max\{|\rho_k^{ij}| \mid 0 \le i, j, k \le n$, then

$$d(st,s_0t_0) \le d(st,s_0t) + d(s_0t,s_0t_0)$$

and, for example, $d(st,s_0t) \le d(s,s_0)d(t,0)n^2M$. We leave it to the reader to write out the details.

4.2. Theorem [Cohen and Collins (1959)]: If T is a compact affine right topological semigroup with identity, then every invertible element of T is an extreme point.

Proof: It suffices to show that the identity 1 is an extreme point since, if s is invertible, then ρ_s is an affine isomorphism of T and maps extreme points onto extreme points. Suppose

$$1 = as + bt, \quad s, t \in T, \ 0 < a, b < 1, \ a + b = 1.$$

For any extreme point $s_0 \in T$, $s_0 = s_0(as + bt) = as_0s + bs_0t$, hence $s_0s = s_0t$. It follows that $t_0s = t_0t$ for all convex combinations t_0 of extreme points. Taking a net of convex combinations of extreme points converging to 1 (Krein-Milman theorem [Dunford and Schwartz (1964); Theorem V.8.4]) and using the continuity of ρ_s and ρ_t, we obtain s = t.

Among the examples which show that the converse of Theorem 4.2 fails are the interval [0,1] with the usual multiplication and the interval [-1,1] with all products equal to zero.

4.3. Theorem: Let T be a compact affine right topological semigroup. Then

(i) the minimal left ideals and minimal (closed) right ideals of T are convex.

(ii) the minimal left (right) ideals of T are left (right) zero semigroups.

(iii) the maximal subgroups of the minimal (two-sided) ideal K(T) are trivial, consisting of an identity only.

Proof: (i) If L is a minimal left ideal of T and s ∈ L, then L = Ts and hence is convex. Similarly, minimal right ideals are convex, and a closure argument shows minimal closed right ideals are convex.

(ii), (iii) Let L be a minimal left ideal of T and let $s \in L$. Since L is closed (Theorem 2.2 (i)) and convex, it follows from any of the usual fixed point theorems [Dunford and Schwartz (1964); V. 10] that ρ_s has a fixed point in L. But $\{t \in T \mid \rho_s t = ts = t\}$ is a left ideal of T contained in L = Ls, hence equals L, and ts = t for all $t \in L$, as required. This and Theorem 2.2 imply (iii); hence, minimal right ideals are right zero semigroups.

4.4. Corollary: A compact affine right topological group consists of an identity element only.

4.5. Corollary: Let T be a compact affine right topological semigroup. Then the closure operation gives a one-to-one correspondence between the minimal right ideals of T and the minimal closed right ideals of T.

Proof: This follows from Theorem 2.3 (iii).

The next few results are concerned with what information can be got about a compact affine semigroup T from assumptions made about a subsemigroup S satisfying $\overline{\text{co}}S = T$. We note that Example V.1.9, which was communicated to us by J. W. Baker, shows that the hypothesis $S \subset \Lambda(T)$ cannot be dropped in the next theorem, and that Theorem 4.1 implies any such example must be infinite dimensional. Note also Example V.1.4, where $T = \text{co}\{0,2\} = [0,2]$, is non-topological, while $S = \{0,2\}$ is a right zero semigroup.

4.6. <u>Theorem</u>: Let T be a compact affine right topological semigroup with a compact topological (semitopological) subsemigroup S such that $T = \overline{co}S$ and $S \subset \Lambda(T) = \{t \in T \mid \lambda_t: T \to T$ is continuous$\}$. Then T is topological (semitopological).

Proof: See Theorems III.8.7 and III.9.7 ahead.

4.7. <u>Theorem</u>: Let T be a compact affine right topological semigroup with a subsemigroup S satisfying $\overline{co}S = T$ and $S \subset \Lambda(T)$. Let $A(S)$ be the set of restrictions to S of the affine functions in $C(T)$. Then $A(S)$ has a left invariant mean if and only if the minimal two-sided ideal $K(T)$ is a minimal right ideal.

Proof: The hypotheses imply that $A(S) = A$ is left introverted and hence that $M(A)$, the set of means on A, is a compact right topological semigroup (Theorem I.4.14). Also, the canonical map $s \to e(s)$ extends to a continuous affine isomorphism of T onto $M(A)$. The desired conclusions now follow from Remark I.4.15 (b) and Theorem 4.3 (ii).

4.8. <u>Remarks</u>: 1. A left introverted subspace of $C(S)$ that contains $A(S)$ (of Theorem 4.7) is the subspace $U(S)$ of $C(S)$ whose members are uniformly continuous with respect to the uniformity S gets as a subspace of the compact uniform space T, which is the same as the uniformity S gets as a subspace of the compact uniform space $\overline{S} \subset T$. Thus the assumption that $U(S)$ has a left invariant mean implies that $A(S)$ has a left invariant mean.

However, the following example shows that, even when T is topological, the fact that K(T) is a minimal right ideal need not imply U(S) has a left invariant mean (LIM). Let T be the convex hull of

$$S = \left\{ \begin{pmatrix} 1 & 0 & 1 \\ 1 & 0 & 1 \\ 0 & 0 & 0 \end{pmatrix}, \begin{pmatrix} -1 & 0 & -1 \\ 1 & 0 & 1 \\ 0 & 0 & 0 \end{pmatrix}, \begin{pmatrix} 1 & 0 & 1 \\ -1 & 0 & -1 \\ 0 & 0 & 0 \end{pmatrix}, \begin{pmatrix} -1 & 0 & -1 \\ -1 & 0 & -1 \\ 0 & 0 & 0 \end{pmatrix} \right\},$$

$$T = \left\{ \begin{pmatrix} a & 0 & a \\ b & 0 & b \\ 0 & 0 & 0 \end{pmatrix} \;\middle|\; -1 \le a,\ b \le 1 \right\}$$

with the usual matrix multiplication. Then the zero matrix is a zero and the minimal ideal for T, while S is (isomorphic to) a left-group and $C(S) = U(S)$ has no left invariant mean. (To get an example with identity, one can add the identity matrix to S.)

2. This example also shows that (in the notation of Theorem 4.7 with S closed in T) K(T) can be a minimal right ideal when K(S) is not. We note that, if T is semitopological and S is not necessarily closed, then K(T) must be a minimal right ideal if $K(\overline{S})$ is a minimal right ideal [Berglund and Hofmann (1967)] (which follows from Theorem 5.6 ahead and the affine analog of Theorem III.8.4). This last conclusion will still hold with T and S as in Theorem 4.7 if one requires the maximal subgroups of $K(\overline{S})$ to be compact topological groups (and proceeds as in Proposition 5.4 ahead).

To get a variant of Theorem 4.7 with $A(S)$ replaced by U(S) (of Remark 4.8, 1) we need a setting where the phenomenon of the example in Remark 4.8, 1, cannot occur. One such setting is that of (Choquet) simplexes; see [Alfsen (1971)] for the definition (page 84) and relevant theory.

4.9. Theorem: Let T be a simplex and a compact affine right topological semigroup with a subsemigroup S such that $S \subset \Lambda(T)$ and $\overline{\mathrm{co}S} = T$. Let exT denote the extreme points of T and suppose $S \subset \overline{exT}$ (closure in T). Assume either

(a) T is metrizable, or

(b) exT is closed.

Then U(S) has a LIM if and only if K(T) is a minimal right ideal.

Proof: By Remark 4.8, 1, and Theorem 4.7, we only need to prove U(S) has a LIM if K(T) is a minimal right ideal. Assume (a) holds. By 4.3 (ii), K(T) is a right zero semigroup. Choose any $t \in K(T)$ and define μ on U(S) by $\mu(f) = \tilde{f}(t)$, where $\tilde{f}$ is the unique affine Borel extension of f to T [Alfsen (1971); Proposition II.3.14]. If $s \in S$, then $L_s\tilde{f} = \tilde{f} \circ \lambda_s$ is an affine Borel extension of L_sf; hence $L_s\tilde{f} = (L_sf)^{\sim}$ and

$$\mu(L_sf) = (L_sf)^{\sim}(t) = \tilde{f}(st) = \tilde{f}(t) = \mu(f),$$

i.e., μ is a LIM on U(S).

If (b) holds, the proof is similar; one uses [Alfsen (1971); Proposition II.3.13].

4.10. Remarks: In the setting of Theorem 4.9 with T also semitopological, one can use Theorem 5.6 ahead to show that K(T) is a minimal right ideal if and only if K(S) is a minimal right ideal. It can be shown that this last assertion can fail to hold if T is not required to be semitopological.

4.11. Theorem: Let T be a compact affine semitopological semigroup with a subgroup G such that $\overline{\mathrm{co}}G = T$. Then the minimal ideal K(T) has only one member, a zero for T.

Proof: This follows from Corollary IV.1.14, Remarks III.8.9 and the universal mapping property Theorem III.8.4 ahead.

4.12. Remarks: 1. Theorem 4.11 generalizes Theorem 2 in [Cohen and Collins (1959)] and is stated along with a further generalization in [Berglund and Hofmann (1967); III.4.18-19]. However, the example in Remarks 4.8, 1, shows that II.3.22 of [Berglund and Hofmann (1967)] can fail and hence the proof of Proposition III.4.16 (on which the proof of III.4.19 depends) has a gap at step (3). This gap can be filled using Theorem 5.6 ahead.

2. The non-uniqueness of invariant mean on *LUC*(G) (see §III.5 ahead) for amenable, non-compact, locally compact groups G (see [Chou (1970)]) shows that the conclusion of Theorem 4.11 can fail if T is assumed to be only right topological. See Remark 5.9 (i) ahead for a related problem.

The next few results are concerned with the possible convexity of the minimal two-sided ideal K(T) of a compact affine right topological semigroup T. Cohen and Collins (1959) characterized all one dimensional affine semigroups (there are five of them) and all two dimensional compact affine semigroups (there are seventeen of them). (Bear in mind Theorem 4.1 at this point: all finite dimensional affine semigroups are topological.) They also show that, if a compact affine semigroup T has dimension less than 3 or is of dimension 3 and has an identity, then K(T) is convex. And they give examples which show these results are best possible; see pp. 111, 112 in their paper.

4.13. Theorem [Collins (1962)]: If T is an affine semigroup with set of idempotents E(T) and L is a line segment in T containing three distinct idempotents, then L consists entirely of idempotents and $fLf = \{f\}$ for all $f \in L$.

Proof: Let $e, f, g \in E(T)$, $a \in (0,1)$ be such that $e = af + (1 - a)g$. Squaring yields, after simplification, $f + g = fg + gf$, which is independent of a; and doing the reverse of the simplification step shows $bf + (1 - b)g \in E(T)$ for any $b \in (0,1)$.

The last assertion of the theorem clearly holds (since $fgf = f$ and $gfg = g$ also hold in the previous paragraph).

4.14. Corollary [Collins (1962)]: Let T be an affine semigroup and consider the following statements about T.

(a) $T = E(T)$.

(b) $sTs = s$ for all $s \in T$.

(c) $T = K(T)$.

Then (a) and (b) are equivalent and imply (c); and (c) implies (a) and (b) at least if T is also compact right topological.

Proof: That (a) implies (b) follows directly from Theorem 4.13 while, if (b) holds and $s \in T$, then $s^3 = s$ and $s^2 = s^3s = ss^2s = s$. It is obvious that (b) implies (c), and that (c) implies (a) in the restricted setting follows from the structure theorem for K(T) (Theorem 2.2) and Theorem 4.3 (ii).

4.15. Corollary: Let T be a compact affine right topological semigroup with minimal ideal K(T). Then K(T) is convex if and only if, for each pair e, f $\in$ K(T), the open line segment between e and f meets K(T).

4.16. Theorem: Let (X,S) be an affine flow with enveloping semigroup E (I.2.1). Suppose that X contains a fixed point z of S. Let K = $K(\overline{co}E)$ be the minimal ideal of the closed convex hull of E. Then the following statements are equivalent:

(a) $\overline{co}E$ contains a right zero.

(b) K is a right zero semigroup.

(c) eX is invariant for some e $\in$ K.

(d) eX is invariant for every e $\in$ K.

(e) eX = fX for all e, f $\in$ K.

(f) eX = X_g for every e $\in$ K.

(g) eX = X_g for some e $\in$ K.

(h) $X_g = X_r = \{x \in X \mid y \in \overline{co}(Sx) \text{ implies } x \in \overline{co}(Sy)\}$.

(i) $X_g = X_p = \{x \in X \mid x \text{ is a fixed point of } S\}$.

(j) $\overline{co}(Sx)$ contains a fixed point of S for every x $\in$ X.

Moreover, $\overline{co}E$ has a zero element if and only if $\overline{co}(Sx)$ contains a unique fixed point of S for every x $\in$ X.

Proof: This follows from 4.3 (iii), 1.37, and the observations that $\overline{co}(Sx) = (\overline{co}E)x$ and that a point w $\in$ X is a fixed point of S if and only if it is a fixed point of $\overline{co}E$.

4.17. Proposition: Let S be a semitopological semigroup, and let F be a translation invariant, left introverted, conjugate closed, norm closed subspace of $C(S)$ containing the constant functions. Suppose (ψ, X) is an F-affine compactification of S (§III.2 ahead). Then the following statements are equivalent:

(a) X contains a right zero.

(b) The minimal ideal $K(X)$ is a right zero semigroup.

(c) $\overline{co}(\psi(S)x)$ contains a right zero for every $x \in X$.

(d) The pointwise closure of $co(R_S f)$ contains a constant function for each $f \in F$.

(e) F is left amenable.

Moreover, if X is right reductive ($ax = bx$ for every $x \in X$ implies that $a = b$), then the above statements are equivalent to

(f) $K(X) = \{x \in X \mid y \in \overline{co}(\psi(S)x) \text{ implies that } x \in \overline{co}(\psi(S)y)\}$.

Also, if X is right reductive, then the following statements are equivalent:

(g) X contains a zero element z.

(h) $\overline{co}(\psi(S)x)$ contains exactly one right zero for every $x \in X$.

(i) F is (left and right) amenable.

5. SUPPORT OF MEANS

In this section means are identified with probability measures and, as such, their supports are analysed. In Remarks 5.9, a number of intriguing examples and unsolved problems in the right topological setting are discussed. Theorems 5.6 and 5.7 show things are much simpler in the semitopological setting.

Let X be a compact Hausdorff space and let P(X) denote the set of all (regular Borel) probability measures on X. If $\mu \in P(X)$, the support of μ, denoted supp μ, is defined to be the intersection of all closed subsets F of X such that $\mu(F) = 1$. (Equivalently, supp μ is the unique closed set $A \subset X$ such that $\mu(A) = 1$ and $\mu(U) > 0$ for each open set $U \subset X$ with $U \cap A \neq \emptyset$.) And, if $B \subset P(X)$, by supp B we mean the closure in X of $\cup\{\text{supp } \mu \mid \mu \in B\}$.

The connection with means is as follows. Let S be a semitopological semigroup and let $\mathcal{F}$ be a translation invariant left introverted C*-subalgebra of $\mathcal{C}(S)$. Then $MM(\mathcal{F})$ is a compact right topological semigroup as is $M(\mathcal{F})$ (Theorem I.4.14), the Gelfand mapping provides an isomorphism of $\mathcal{F}$ and $C(MM(\mathcal{F}))$ and, by the Riesz representation theorem, we may identify $M(\mathcal{F})$ with $P(MM(\mathcal{F}))$ via an affine isomorphism (much as in Definition I.1.10 and Theorem I.4.16); thus by the support of a mean $\mu \in M(\mathcal{F})$ we are referring to the support in $MM(\mathcal{F})$ of the associated probability measure in $P(MM(\mathcal{F}))$, which we also denote by μ. We use the notation (as in I.3.4) that e is the map of S into $MM(\mathcal{F})$ (or $M(\mathcal{F})$) defined by $e(s)f = f(s)$ for all $f \in \mathcal{F}$; and if $A \subset M(\mathcal{F})$, $\overline{\text{co}}A$ denotes the closed convex hull in $M(F)$ of A. A subset B of $M(F)$ is called <u>extremal</u> (in $M(F)$) if B is compact and convex and every open segment in $M(F)$ that contains a point of B lies entirely in B. (Of course, the definition works for any compact convex set.)

The first three results here appear (essentially) in [Wilde and Witz (1967)].

5.1. <u>Lemma</u>: If $\mu \in M(F)$, then supp μ is the smallest closed subset A of $MM(F)$ such that $\mu \in \overline{co}A$.

Proof: This follows from the fact that, for closed $A \subset MM(F)$, $\mu(A) = 1$ if and only if $\mu \in \overline{co}A$.

5.2. <u>Theorem</u>: If A is a closed subset of $MM(F)$, then $A = \text{supp}(\overline{co}A)$. If B is an extremal subset of $M(F)$ then $B = \overline{co}(\text{supp } B)$. The mapping $A \to \overline{co}A$ is a one-to-one correspondence between the closed subsets of $MM(F)$ and the extremal subsets of $M(F)$.

Proof: The first assertion is obvious. Suppose B is an extremal subset of $M(F)$. Clearly $B \subset \overline{co}(\text{supp } B)$ and, by the Krein-Milman theorem, B is the closed convex hull of its extreme points, each of which is an extreme point of $M(F)$, hence in $MM(F)$, since B is extremal in $M(F)$. It follows that $B \supset \overline{co}(\text{supp } B)$.

It remains to show that $\overline{co}A$ is extremal in $M(F)$ when A is a closed subset of $MM(F)$. If $\mu \in \overline{co}A$, $\mu = a\mu_1 + (1 - a)\mu_2$ with $\mu_1, \mu_2 \in M(F)$ and $a \in (0,1)$, then $\mu(A) = 1 = \mu_1(A) = \mu_2(A)$. Hence supp $\mu_i \subset A$ and $\mu_i \in \overline{co}A$, $i = 1,2$, by Lemma 5.1.

5.3. Theorem [Wilde and Witz (1967), Fairchild (1972)]: If A is a closed left ideal of $MM(F)$, then $\overline{co}A$ is a left ideal of $M(F)$. If μ is a left invariant mean (LIM) on F, then supp μ is a (closed) left ideal in $MM(F)$.

Proof: The first statement follows from the fact that $e(S) \subset \Lambda(M(F)) = \{\mu \in M(F) \mid \lambda_\mu : M(F) \to M(F) \text{ is continuous}\}$ (Theorem I.4.14). As to the second statement, suppose μ is a LIM on F, $\nu \in$ supp μ and $\nu' \in MM(F)$ with $\nu'\nu \notin$ supp μ. By the continuity of $\rho_\nu : MM(F) \to MM(F)$ and the fact that supp μ is closed, there is a $t \in S$ with $e(t)\nu \notin$ supp μ. Also, there is an $f \in F$ with $0 \leq f \leq 1$ and $\mu'(f) = 0$ for all $\mu' \in$ supp μ and hence $\mu(f) = 0$ (Lemma 5.1), while $e(t)\nu(f) = \nu(L_t f) = 1$, which implies $\mu(L_t f) > 0$ and contradicts the hypothesis that μ is left invariant.

If F is left amenable, then LIM$(F) \neq \emptyset$ and is a compact (in the weak * topology) convex subset of F^* and hence is the closed convex hull of its extreme points, which we call extreme LIM's; also, a mean $\mu \in M(F)$ is a LIM if and only if, when regarded as a probability measure on $MM(F)$, it is invariant in the sense that

$$\mu(A) = \mu(e(s)^{-1}A) = \mu(\{\nu \in MM(F) \mid e(s)\nu \in A\})$$

for all $s \in S$ and all Borel subsets $A \subset MM(F)$. An obvious way to get an extreme LIM is as follows.

5.4. Proposition: Suppose the minimal left ideals of $MM(F)$ are compact topological groups and H is one such group. Then the mean in $M(F)$ that takes a member of F and integrates the restriction of it to H with respect to (normalized) Haar measure on H is an extreme LIM.

Proof: The result follows from the uniqueness of Haar measure.

In Proposition 5.4 the mean μ is the only LIM supported on the minimal left ideal H; see Remark 5.9 (ii), ahead, in this regard.

5.5. Proposition [Wilde and Witz (1967)]: If F is left amenable and A is a closed left ideal in MM(F), then $\overline{\text{co}}A \subset M(F)$ contains an extreme LIM, i.e., A supports an extreme LIM.

Proof: By Theorem 5.3, $\overline{\text{co}}A$ is a left ideal of M(F), hence meets LIM(F), and, by Theorem 5.2, $\overline{\text{co}}A$ is an extremal subset of M(F), which implies that $\overline{\text{co}}A \cap \text{LIM}(F)$ is an extremal subset of the compact convex set LIM(F) and thus contains an extreme LIM.

One might expect that the support of an extreme LIM would be a minimal left ideal of MM(F). This turns out not to be the case in general; see Remark 5.9 (i), ahead. However, it is the case if MM(F) is a semitopological semigroup (i.e., $F \subset \mathcal{WAP}(S)$ (Remark III.8.6 (c))). We proceed to prove this. The next result is the first step and is related to Theorem 4.9; see Remarks 4.10.

5.6. Theorem [deLeeuw and Glicksberg (1961), Berglund and Hofmann (1967); p. 83]: Suppose MM(F) is a semitopological semigroup. Then F is left amenable if and only if the minimal ideal K(MM(F)) is a right-group. Thus, the minimal ideal of

$M(F)$ is a minimal right ideal if and only if the minimal ideal of $MM(F)$ is a minimal right ideal.

Proof: If $K(MM(F))$ is a minimal right ideal, i.e., is a right-group, then the minimal left ideals are compact topological groups (Remark 2.11) and Proposition 5.4 and Remark I.4.15 (b), complete the proof in one direction. And, if $K(MM(F))$ contains two minimal right ideals R_1 and R_2, they are closed and there is a function $f \in F$, $0 \leq f \leq 1$, with $\nu(f) = 0$ if $\nu \in R_1$ and $\nu(f) = 1$ if $\nu \in R_2$. Picking $\nu_1 \in R_1$, $\nu_2 \in R_2$ and working in the semigroup $MM(F)$, we have $L_{\nu_1}\hat{f} = 0$, $L_{\nu_2}\hat{f} = 1$ (where $\hat{f} \in C(MM(F))$ satisfies $\hat{f}(e(s)) = e(s)f = f(s)$, $s \in S$), which implies $C(MM(F))$ is not left amenable; hence F is not left amenable. (See Theorem I.4.16 and Remark I.4.15 (b), in this regard.)

5.7. <u>Theorem</u> [Berglund and Hofmann (1967)]: Suppose F is left amenable and $MM(F)$ is a semitopological semigroup. Then the support of a LIM μ on F is a right-group contained in the minimal ideal $K(MM(F))$, and μ is an extreme LIM if and only if its support is a minimal left ideal of $MM(F)$, i.e., it is of the form described in Proposition 5.4.

Proof: Since F is left amenable, F has a LIM, the minimal ideal of $M(F)$ is a minimal right ideal and the minimal ideal K of $MM(F)$ is a minimal right ideal (Theorem 5.6). Hence K is topologically isomorphic to the right-group $G \times E$ via

$$(s,f) \to sf: \quad G \times E \to K,$$

where $G = Ke$ and $e \in E(K) = E$ (Remark 2.11).

Now, let μ be a LIM on F. The regularity of μ as a probability measure on $MM(F)$, together with its left invariance, imply supp $\mu \subset K$. Thus we may consider μ as a probability measure on K. For a fixed non-negative $g \in C(E)$ define

$$\nu_g(h) = \mu(h \otimes g),\ h \in C(G),$$

where $h \otimes g \in C(G \times E)$ is defined by

$$(h \otimes g)(s,f) = h(s)g(f).$$

Then $\nu_g(L_t h) = \nu_g(h)$ for $t \in G$, so ν_g must be a multiple of normalized Haar measure ν on G. Thus, there exists $\lambda(g) \geq 0$ such that

$$\mu(h \otimes g) = \lambda(g)\nu(h).$$

This equation is meaningful for arbitrary $g \in C(E)$ and defines a mean λ on $C(E)$; in fact $\lambda(g) = \mu(1 \otimes g)$. Therefore supp μ is the right-group (and left ideal) $G \times$ supp λ.

To prove the second part of the theorem, suppose supp μ is a minimal left ideal of $MM(F)$, i.e., supp λ is a singleton. Then μ is extreme, as in Proposition 5.4. If supp μ is not a minimal left ideal of $MM(F)$, then supp λ contains more than one point and we can write supp λ = $V_1 \cup V_2$, where V_1 is an open neighbourhood of one point of supp λ and $V_2 = V_1'$ is a neighbourhood of another point of supp λ. It follows that $\lambda(V_i) \neq 0$ and we can define a probability measure λ_i on E by

$$\lambda_i(A) = \lambda(A \cap V_i)/\lambda(V_i)$$

for Borel subsets $A \subset E$, $i = 1, 2$, and have

$$\lambda = \lambda(V_1)\lambda_1 + \lambda(V_2)\lambda_2,$$

i.e., λ is not extreme.

5.8. Corollary [deLeeuw and Glicksberg (1961)]: Suppose $MM(F)$ is a semitopological semigroup. Then F is (left and right) amenable if and only if $K(MM(F))$ is a compact topological group. In this event, there is precisely one (left and/or right) invariant mean on F.

Proof: This follows mainly from Theorems 5.6 and 5.7.

5.9. Remarks: A number of the conclusions of the last few results fail if $MM(F)$ is allowed to be merely right topological.

(i) Chou (1969) proved that F has an extreme LIM whose support is not a minimal left ideal (of $MM(F)$) when S is the (discrete, additive) semigroup of natural numbers and $F = C(S) = LMC(S)$. (See III.4 ahead for the definition of $LMC(S)$.) Fairchild (1972) proved that, if S is a discrete semigroup, then $C(S)$ has an extreme LIM whose support is not a minimal left ideal of $MM(C(S))$ if and only if S has a subset A satisfying:

$$\mu(\chi_A) > 0 \text{ for some LIM } \mu \text{ on } C(S), \text{ but}$$
$$\mu'(\chi_{K^{-1}A}) < 1 \text{ for every LIM } \mu' \text{ and every finite } K \subset S.$$

She also showed that all infinite solvable groups and all countably infinite, locally finite groups have such a subset.

(ii) The key to the proof of one part of Theorem 5.6 is the fact that minimal right ideals in a compact semitopological semigroup are closed. In a compact right topological semigroup, the minimal right ideals need not be closed (Examples V.1, 3 and 10), and one can use V.1.10 to show that, for any infinite discrete abelian group G, the minimal ideal of MM(C(G)) is not a minimal right ideal, while C(G) has a LIM (since G is abelian) and hence the minimal ideal of M(C(G)) is a minimal right ideal (see Remark I.4.15 (b)).

A further complication in the right topological setting was discovered by Raimi (1964). He showed that, if S is the semigroup of natural numbers and $F = C(S)$, then every minimal left ideal of MM(F) = βS supports more than one extreme LIM. (His methods also work for the group of integers.) And Chou (1971) showed that, if G is any countable infinite discrete amenable group and L is a closed left ideal in βG, then the (compact, convex) subset of LIM(C(G)) supported on L is infinite dimensional.

We make the related observation that, although (normalized) Haar measure on a compact topological group is unique, it is unlikely that the invariant measure on a compact right topological group G (Theorem 3.2) is unique. The set of such measures is a compact convex subset of P(G) and is the closed convex hull of its extreme points; if $\Lambda(G)$ is dense in G (as in the enveloping group of Example V.1.7), the support of any such measure is G whether the measure is extreme or not.

(iii) Let G be the group of integers and let us consider the flow (βG,G) (as in Remarks I.4.4):

$$(\mu, s) \to L_s^*\mu = e(s)\mu, \quad \mu \in \beta G,\ s \in G.$$

Let ν_1, ν_2 be distinct extreme LIM's on C(G) supported on the same minimal left ideal $L \subset \beta S$ (as in the previous remark). It follows from [Blum and Hanson (1960); Corollary 2] that there are disjoint invariant Borel subsets A_1 and A_2 of L with $\nu_1(A_1) = 1$, $\nu_2(A_2) = 1$ and $\bar{A}_1 = L = \bar{A}_2$, where a subset A of βG is <u>invariant</u> if

$$e(s)\mu \in A, \quad s \in G, \ \mu \in \beta G.$$

(See also [Phelps (1966); Chapter 10] in this regard.)

(iv) Results related to or extending Theorem 5.7 are given in [Pym (1968), Mukherjea and Tserpes (1973)].

(v) Invariant means and their supports in a somewhat different context are considered by Rosenblatt (1976).

(vi) The conclusion of Corollary 5.8 fails if $MM(F)$ is allowed to be merely right topological; Chou (1970) has shown that, for amenable, σ-compact, unimodular, locally compact groups G, the linear span of the set of invariant means on $LUC(G)$ is infinite dimensional. (See III.5 ahead for the definition of $LUC(G)$.)

We end this section with some results in the semitopological setting that are due to Glicksberg (1961). They are related to earlier results of this section, and for them we recall that if $MM(F)$ is a compact semitopological semigroup, then we may identify $M(F)$ algebraically and topologically with $P(MM(F))$ (I.4.16). We shall also identify F with $C(MM(F))$. Thus, under these identifications,

$$\mu(f) = \int f(x)\mu(dx), \quad f \in F, \ \mu \in M(F).$$

5.10. <u>Lemma</u>: Let $MM(F)$ be semitopological. If $\mu, \nu \in M(F)$, then

$$\text{supp } \mu\nu = [(\text{supp } \mu) \cdot (\text{supp } \nu)]^-.$$

Proof: Let $A = \text{supp } \mu$, $B = \text{supp } \mu$ and $C = \overline{AB}$. Then, since C is compact and $\mu\nu$ is regular, there exists for each $\varepsilon > 0$ an open subset U of $X = MM(F)$ containing C such that $(\mu\nu)(U) \leq (\mu\nu)(C) + \varepsilon$. Choose $f \in F$ such that $0 \leq f \leq 1$, $f(C) = 1$, and $f(U') = 0$. Then, for all $x, y \in X$, $\chi_A(x)\chi_B(y) \leq \chi_C(xy) \leq f(xy)$, so

$$1 = \mu(A)\nu(B) = \iint \chi_A(x)\chi_B(y)\nu(dy)\mu(dx) \leq \iint f(xy)\nu(dy)\mu(dx)$$

$$= \int_U f(z)\mu\nu(dz) \leq \mu\nu(U) \leq \mu\nu(C) + \varepsilon.$$

Therefore $\mu\nu(C) = 1$, and $\text{supp } \mu\nu \subset C$.

To prove the reverse inclusion it suffices to show that, if W is open in X and $W \cap C \neq \emptyset$, then $\mu\nu(W) > 0$. Now, if W meets C then there exist $x_0 \in A$, $y_0 \in B$ such that $x_0y_0 \in W$. Choose $g \in F$ such that

$$0 \leq g \leq 1,\ g(x_0y_0) = 1,\ g(W') = 0.$$

Then $y \to g(x_0y)$ is positive on some neighbourhood of y_0, hence

$$\int g(x_0y)\nu(dy) > 0.$$

Since the function $x \to \int g(xy)\nu(dy)$ is continuous (I.1.8 (b)), it is positive on some neighbourhood of x_0, and therefore

$$0 < \iint g(xy)\nu(dy)\mu(dx) = \mu\nu(g) \leq \mu\nu(W).$$

5.11. <u>Definition</u>: A semitopological semigroup is called <u>topologically simple</u> if it contains no proper closed ideals.

5.12. <u>Theorem</u>: Suppose $X = MM(F)$ is semitopological and μ is an idempotent in $M(F)$. Then

(a) $Y = \text{supp}\,\mu$ is a topologically simple subsemigroup of X;

(b) if $C(Y)$ is left amenable, μ is left invariant on Y, and Y is a topological right-group;

(c) if $C(Y)$ is amenable, Y is a topological group and μ is Haar measure on Y. In particular, this is the case if X is either commutative or a group.

Proof: That Y is a semigroup follows immediately from Lemma 5.10. Let I be any closed ideal of Y. To prove $I = Y$ we show first that for any real-valued $f \in F$ there exists a minimal right ideal $J(f)$ of Y such that for all $x \in J(f)$,

$$\mu(L_x f) = \sup_{y \in Y} \mu(L_y f).$$

Choose $y_0 \in Y$ such that $\mu(L_{y_0} f) = \sup_{y \in Y} \mu(L_y f)$. Then

$$\mu(L_{y_0} f) = \mu^2(L_{y_0} f) = \iint f(y_0 yz)\mu(dz)\mu(dy)$$

$$= \int \mu(L_{y_0 y} f)\mu(dy) \leq \mu(L_{y_0} f)$$

since $\mu(L_z f) \leq \mu(L_{y_0} f)$ for $z \in y_0 Y \subset Y$. Thus

$$\mu(L_{y_0} f) = \int \mu(L_{y_0 y} f)\mu(dy),$$

and this implies that $\mu(L_{y_0} f) = \mu(L_{y_0 y} f)$ for all $y \in Y$.

Therefore, if $J(f)$ is any minimal right ideal contained in $y_0 Y$, $\mu(L_x f) = \mu(L_{y_0} f)$ for all $x \in J(f)$.

Now suppose $I \neq Y$. Choose non-negative $f \in F$ such that $f(I) = 0$ and f does not vanish identically on Y. Since $(Y \cdot Y)^- = Y$, there exists $y_1 \in Y$ such that $L_{y_1} f$ is not identically zero on Y, and therefore

$$\mu(L_x f) \geq \mu(L_{y_1} f) > 0, \quad x \in J(f).$$

On the other hand if $x \in J(f)$ then $xY \subset I$ and so

$$\mu(L_x f) = \int_Y f(xy)\mu(dy) = 0.$$

This is the desired contradiction and completes the proof of (a).

If $C(Y)$ is left amenable, then $Y = J(f)$ (Theorem 5.6) and hence $\mu(L_x f)$ is constant in $x \in Y$ for each real-valued $f \in F$. Also, if $e^2 = e \in Y$ then $L_e f = f$ on $Y = eY$. Therefore $\mu(L_x f) = \mu(f)$ for all $x \in Y$. That Y is a topological right-group follows from Theorem 5.6 and Remark 2.11.

If $C(Y)$ is amenable (which is the case if Y is commutative) then Y is a group by Corollary 5.8. If X is a group then Y is a group by Theorem 2.2. The rest of (c) follows from (b).

5.13. Remark: The conclusions of (b) and (c) need not hold if $C(Y)$ is not left amenable. For example, let $F = C[0,1]$, where $[0,1] = MM(F)$ has the multiplication $xy = x$. Then $M(F)$ is a (non-trivial) left-zero semigroup so F is not left amenable. If μ is Lebesgue measure on $[0,1]$, then $\mu^2 = \mu$ in $M(F)$, but μ is not left invariant on supp $\mu = [0,1]$.

CHAPTER III

SUBSPACES OF $\mathcal{C}(S)$ AND COMPACTIFICATIONS OF S

In this chapter we show how compact (sometimes affine) right topological semigroups of various kinds can be got as compactifications of S from various subspaces of $\mathcal{C}(S)$, where S is a semitopological semigroup. The techniques used are established in §§ 1 and 2 and are similar to those used to obtain the {weakly} almost periodic compactification from the {weakly} almost periodic functions; and the compactifications have universal mapping properties (i.e., are maximal with respect to certain properties) analogous to that of the {weakly} almost periodic compactification (which is given here in Theorem {8.4} 9.4 ahead). But, see Example V.1.11.

Remarks: (i) These matters are treated very generally from a category point of view in an appendix.

(ii) Definitions of, and notation for, the subspaces of $\mathcal{C}(S)$ considered in §§ 3 - 13 ahead are collected all in one spot at the beginning of §14 and in Appendix B (as well as being given when needed in the course of these sections).

In § 14 we consider what inclusion relationships hold among the subspaces of §§ 3 - 13 and under what conditions various subspaces can be identified, while, in § 15 we ask when a function of a certain type on a subsemigroup S of a semitopological semigroup T extends to a function of the same type on T. The structure theory of Chapter II is used in § 16 to generalize some results of deLeeuw and Glicksberg (1961) on the splitting of $\mathcal{WAP}(S)$.

1. GENERAL THEORY OF AFFINE COMPACTIFICATIONS

Throughout this section F denotes a translation invariant, left introverted, conjugate closed, norm closed, linear subspace of $C(S)$ containing the constant functions (where, of course, S denotes a semitopological semigroup).

1.1. Definition: An F-affine compactification of S is a pair (ψ, X), where X is a compact affine right topological semigroup and $\psi: S \to X$ is a continuous homomorphism with the following properties:

1. $\overline{\mathrm{co}}\,\psi(S) = X$.
2. $\lambda_{\psi(s)}: X \to X$ is continuous for each $s \in S$.
3. $\psi^* A(X) = F$.

Here $A(X)$ denotes the subspace of $C(X)$ consisting of all affine functions and $\psi^*: C(X) \to C(S)$ is the adjoint of ψ. Note that $\psi^*|_{A(X)}$ is an isometry, by 1.

1.2. Remark: F-affine compactifications always exist: take $X = M(F)$, the space of means on F, and $\psi = e$, the canonical injection of S into $M(F)$ (Corollary I.3.7 and Theorem I.4.14). We shall call $(e, M(F))$ the canonical F-affine compactification of S; Corollary 1.6 ahead asserts that this is (up to isomorphism) the only F-affine compactification of S.

1.3. Definition: Let P be a set of properties that pairs of the form (ψ, X) may or may not possess, where X is a compact affine right topological semigroup and $\psi: S \to X$ is a continuous homomorphism. We shall say that (ψ, X) is maximal with respect to P (more briefly, maximal w.r.t. P) if

1. (ψ, X) possesses properties P, and
2. whenever (ψ_1, X_1) possesses properties P then there exists a continuous affine homomorphism $\phi: X \to X_1$ such that the diagram

$$\begin{array}{ccc} & X & \\ {\scriptstyle\psi} \nearrow & & \searrow {\scriptstyle\phi} \\ S & \xrightarrow[\psi_1]{} & X_1 \end{array}$$

commutes.

1.4. <u>Theorem</u>: Let X be a compact affine right topological semigroup and let $\psi: S \to X$ be a continuous homomorphism. Then (ψ, X) is an F-affine compactification of S if and only if it is maximal w.r.t. the following set of properties:

(a) $\overline{co}\psi(S) = X$.

(b) $\lambda_{\psi(s)}: X \to X$ is continuous for all $s \in S$.

(c) $\psi^* A(X) \subset F$.

Proof: Let (ψ, X) be an F-affine compactification. Then, by definition, (ψ, X) has properties (a), (b) and (c). Let (ψ_1, X_1) also have properties (a), (b) and (c). Define $\phi: co\psi(S) \to X_1$ by

$$\phi\left(\sum_{s \in S} a(s)\psi(s)\right) = \sum_{s \in S} a(s)\psi_1(s).$$

ϕ is well defined: Suppose

$$\sum_{s \in S} a(s)\psi_1(s) \neq \sum_{s \in S} b(s)\psi_1(s).$$

Then there exists an $h \in A(X_1)$ such that

$$h\left(\sum_{s} a(s)\psi_1(s)\right) \neq h\left(\sum_{s} b(s)\psi_1(s)\right).$$

By assumption, $f = h \circ \psi_1 \in F$, and we have

$$\sum_s a(s)f(s) \neq \sum_s b(s)f(s).$$

Since (ψ,X) is an F-affine compactification, $f = g \circ \psi$ for some $g \in A(X)$. Then

$$g\left(\sum_s a(s)\psi(s)\right) \neq g\left(\sum_s b(s)\psi(s)\right)$$

and therefore $\sum_s a(s)\psi(s) \neq \sum_s b(s)\psi(s)$.

ϕ is uniformly continuous: Since the unique uniformity on X_1 (X) is given by $A(X_1)$ $(A(X))$, it suffices to show that, for any $h \in A(X_1)$, $h \circ \phi$ is the restriction to $\text{co}\psi(S)$ of a member of $A(X)$. But this follows from Definition 1.3 and the assumption that (ψ_1,X_1) has property (c).

Thus, by property (a), ϕ has a continuous affine extension to X, and $\phi \circ \psi = \psi_1$. Clearly, ϕ is a homomorphism of the semigroup $\text{co}\psi(S)$. Since (ψ,X) and (ψ_1,X_1) satisfy (b) and since X and X_1 are right topological, ψ is, in fact, a homomorphism of X. Therefore (ψ,X) is maximal w.r.t. (a), (b) and (c).

Conversely, assume that (ψ,X) is maximal w.r.t. properties (a), (b) and (c). We must show that $F \subset \psi^*(A(X))$. Let (ψ_1,X_1) be any F-affine compactification of S. Then (ψ_1,X_1) has properties (a), (b) and (c); hence, by assumption, there is a continuous affine homomorphism $\phi: X \to X_1$ such that $\phi \circ \psi = \psi_1$ and, for any $f \in F = \psi_1^*(A(X_1))$,

$$f = \psi_1^*(g) = \psi^*(g \circ \phi) \in \psi^*(A(X)),$$

where $g \in A(X_1)$.

1.5. Corollary: Let S_1 and S_2 be semitopological semigroups. Let $\mathcal{F}_i \subset C(S_i)$ be a translation invariant, left introverted, conjugate closed, norm closed, linear subspace containing the constant functions, and let (ψ_i, X_i) be an $\mathcal{F}_i$-affine compactification of S_i, $i = 1, 2$. If $\phi: S_1 \to S_2$ is a continuous homomorphism, then $\phi^*\mathcal{F}_2 \subset \mathcal{F}_1$ if and only if there is a continuous affine homomorphism $\phi': X_1 \to X_2$ such that the following diagram commutes:

$$\begin{array}{ccccc} & X_1 & \overset{\phi'}{\to} & X_2 & \\ \psi_1 & \uparrow & & \uparrow & \psi_2 \\ & S_1 & \underset{\phi}{\to} & S_2 & \end{array}$$

Proof: Suppose $\phi^*\mathcal{F}_2 \subset \mathcal{F}_1$, $\psi = \psi_2 \circ \phi$ and $X = \overline{co}\psi(S_1)$. Then (ψ, X) has properties (a)-(c) of 1.4, so there exists a continuous homomorphism $\phi': X_1 \to X \subset X_2$ such that $\phi' \circ \psi_1 = \psi$.

On the other hand, if ϕ' exists, let $f \in \mathcal{F}_2$. Then there is an $h \in A(X_2)$ such that $h \circ \psi_2 = f$; hence

$$f \circ \phi = h \circ \psi_2 \circ \phi = h \circ \phi' \circ \psi_1 \in \psi_1^*(A(X_1)) = \mathcal{F}_1.$$

The following uniqueness property for $\mathcal{F}$-affine compactifications is now immediate.

1.6. Corollary: If (ψ_1, X_1) and (ψ_2, X_2) are $\mathcal{F}$-affine compactifications of S, then there exists an affine isomorphism and homeomorphism ϕ of X_1 onto X_2 such that $\phi \bullet \psi_1 = \psi_2$.

For our next result we remind the reader that p denotes the topology of pointwise convergence in $C(S)$ and u the uniform or norm topology.

1.7. Corollary: Let $\mathcal{T}$ denote a family of locally convex topologies on $C(S)$ such that, for each $\tau \in \mathcal{T}$, $p \leq \tau \leq u$, $L_t: C(S) \to C(S)$ is τ-continuous for all $t \in S$, and $f \to f^*: C(S) \to C(S)$ is τ-continuous. Let

$\mathcal{F} = \{f \in C(S) \mid \operatorname{co}R_S f$ is relatively τ-compact for all $\tau \in \mathcal{T}\}$.

Then S has an $\mathcal{F}$-affine compactification. Moreover, if X is a compact affine right topological semigroup and $\psi: S \to X$ a continuous homomorphism, then (ψ, X) is an $\mathcal{F}$-affine compactification if and only if it is maximal w.r.t. the following set of properties:

(a) $\overline{\operatorname{co}}\psi(S) = X$.

(b) $\lambda_{\psi(s)}: X \to X$ is continuous for all $s \in S$.

(c') $x \to T_x h: X \to C(S)$ is τ-continuous for all $\tau \in \mathcal{T}$ and $h \in A(X)$, where $T_x h \in C(S)$ is defined by

$$(T_x h)(s) = h(\psi(s)x), \quad s \in S.$$

Proof: By Lemma I.4.17, $\mathcal{F}$ is a translation invariant, left introverted, conjugate closed, norm closed, linear subspace of $C(S)$ containing 1. Hence S has an $\mathcal{F}$-affine compactification.

For the remainder of the corollary it suffices to prove that if (a) and (b) hold, then property (c') is equivalent to property (c) of 1.4. To this end we note first that

(1) $$T_{\operatorname{co}\psi(S)}h = \operatorname{co}R_S\psi^*h, \quad h \in A(X).$$

Now assume that $x \to T_x h: X \to C(S)$ is τ-continuous for all $\tau \in \mathcal{T}$ and $h \in A(X)$. Then by (1), $\operatorname{co}R_S\psi^*h$ is contained in the τ-compact set $T_X h$, hence $\psi^*h \in \mathcal{F}$. Therefore $\psi^*A(X) \subset \mathcal{F}$.

Conversely, assume $\psi^*A(X) \subset F$, and let $h \in A(X)$. By (1) and the p-continuity of the map $x \to T_x h$, $T_x h$ is contained in the p-closure K of $\mathrm{co}R_S\psi^*h$. Since $\psi^*h \in F$, K is τ-compact and the p and τ topologies agree on K for each $\tau \in T$. Therefore $x \to T_x h$ is τ-continuous for each $\tau \in T$.

The following result is (in view of Theorem 3.2 and Remark 14.5 (i) ahead) a generalization of the theorem (asserted in Remark IV.1.10 (a)) that, for any commutative semigroup S, $\mathcal{B}(S)$ is amenable; it could also be proved using this result about commutative semigroups and Theorem 3.5 ahead.

1.8. Proposition: Let S be a commutative semigroup and let (ψ,X) be an F-affine compactification of S. Then the minimal ideal K(X) consists of right zeros of X.

Proof: Note that $\mathrm{co}\psi(S) \subset \Lambda(X)$ is a dense commutative subset of (the compact affine right topological semigroup) X. Since each maximal subgroup in K(X) is trivial (Theorem II.4.3 (iii)) and, therefore, obviously closed, the result follows from Proposition II.2.6 and Theorem II.4.3 (ii).

1.9. Remark: If the semitopological semigroup S has an identity, then, in the language of the Appendix, conditions 1 and 2 of Definition 1.1 imply that X is a compact affine separate right topological S-module (i.e., an object in the category C A Sep RT S-Mod) under the action $s \cdot x = \psi(s)x$.

2. GENERAL THEORY OF NON-AFFINE COMPACTIFICATIONS

The results of this section are completely analogous to those of the previous section, hence all proofs are omitted.

Throughout this section, F denotes a translation invariant, left m-introverted C*-subalgebra of $C(S)$ containing the constant functions.

2.1. Definition: An F-compactification of S is a pair (ψ, X), where X is a compact right topological semigroup and $\psi: S \to X$ is a continuous homomorphism with the following properties:

1. $\overline{\psi(S)} = X$.
2. $\lambda_{\psi(s)}: X \to X$ is continuous for each $s \in S$.
3. $\psi^* C(X) = F$.

2.2. Remarks: (i) F-compactifications always exist: Take $X = MM(F)$ and $\psi = e$ (cf. I.4.14). We shall call $(e, MM(F))$ the canonical F-compactification of S.

(ii) Since a translation invariant, left introverted C*-subalgebra of $C(S)$ is also left m-introverted, every such subalgebra has an associated compactification (as well as the affine compactification of the previous section).

2.3. Definition: Let P be a set of properties that pairs of the form (ψ, X) may or may not possess, where X is a compact right topological semigroup and $\psi: S \to X$ is a continuous homomorphism. We shall say that (ψ, X) is maximal w.r.t. P if

1. (ψ, X) possesses properties P, and

2. whenever (ψ_1, X_1) possesses properties P then there exists a continuous homomorphism $\phi: X \to X_1$ such that the following diagram commutes:

$$\begin{array}{ccc} & X & \\ \psi \nearrow & & \searrow \phi \\ S & \xrightarrow[\psi_1]{} & X_1 \end{array}$$

2.4. <u>Theorem</u>: Let X be a compact right topological semigroup and let $\psi: S \to X$ be a continuous homomorphism. Then (ψ, X) is an F-compactification of S if and only if it is maximal w.r.t. the following set of properties:

(a) $\overline{\psi(S)} = X$.

(b) $\lambda_{\psi(s)}: X \to X$ is continuous for all $s \in S$.

(c) $\psi^* C(X) \subset F$.

2.5. <u>Corollary</u>: Let S_1 , S_2 be semitopological semigroups. Let $F_i \subset C(S_i)$ be a translation invariant, left m-introverted C*-subalgebra containing 1, and let (ψ_i, X_i) be an F_i-compactification of S_i , i = 1, 2. If $\phi: S_1 \to S_2$ is a continuous homomorphism, then $\phi^* F_2 \subset F_1$ if and only if there exists a continuous homomorphism $\phi': X_1 \to X_2$ such that the following diagram commutes:

$$\begin{array}{ccc} X_1 & \xrightarrow{\phi'} & X_2 \\ \psi_1 \uparrow & & \uparrow \psi_2 \\ S_1 & \xrightarrow[\phi]{} & S_2 \end{array}$$

As a consequence of Corollary 2.5 we have the following uniqueness property for F-compactifications.

2.6. Corollary: If (ψ_1, X_1) and (ψ_2, X_2) are F-compactifications of S, then there exists an isomorphism and homeomorphism ϕ of X_1 onto X_2 such that $\phi \circ \psi_1 = \psi_2$.

2.7. Corollary: Let T denote a family of locally convex topologies τ on $C(S)$ such that, for each $\tau \in T$, $p \leq \tau \leq u$, $L_t: C(S) \to C(S)$ is τ-continuous for all $t \in S$ and $f \to f^*$: $C(S) \to C(S)$ is τ-continuous. Let

$F = \{f \in C(S) \mid R_S f$ is relatively τ-compact for all $\tau \in T\}$,

and suppose F is closed under multiplication. Then S has an F-compactification. Furthermore, if X is a compact right topological semigroup and ψ: S $\to$ X is a continuous homomorphism, then (ψ, X) is an F-compactification if and only if it is maximal w.r.t. the following set of properties:

(a) $\overline{\psi(S)} = X$.

(b) $\lambda_{\psi(s)}: X \to X$ is continuous for all $s \in S$.

(c) $x \to T_x h: X \to C(S)$ is τ-continuous for all $\tau \in T$ and $h \in C(X)$.

2.8. Remarks: (i) A typical application of 2.5 is the following: Let τ_i be a locally convex topology on $C(S_i)$ with the properties stated in 2.7, and suppose

$F_i = \{f \in C(S_i) \mid R_S f$ is relatively τ_i-compact$\}$ is closed under multiplication, i = 1, 2. If $\phi: S_1 \to S_2$ is a continuous homomorphism with the property that $\phi^*: C(S_2) \to C(S_1)$ is $\tau_2 - \tau_1$ continuous, then the mapping $\phi': X_1 \to X_2$ of 2.5 exists. This follows immediately from 2.5 and the observation that $R_s \phi^* f = \phi^*(R_{\phi(s)} f)$, $s \in S_1$, $f \in F_2$.

A similar remark holds in the affine case.

(ii) A related remark is that, if S_i is a semitopological semigroup and $F(S_i)$ is one of the subspaces defined in §§ 3 - 13 ahead, i = 1, 2, and if $\phi: S_1 \to S_2$ is a continuous homomorphism, then the adjoint ϕ^* maps $F(S_2)$ into $F(S_1)$.

(iii) If S has an identity and (ψ, X) is an F-compactification of S, then, in the language of the appendix, X is a compact separate right topological S-module (i.e., an object in C Sep RT S-Mod) under the action $s \cdot x = \psi(s)x$.

3. THE $\mathcal{WLUC}$-AFFINE COMPACTIFICATION

3.1. Definition: $\mathcal{WLUC}(S) = \mathcal{WLUC} = \{f \in C(S) \mid s \to L_s f$ is $\sigma(C(S), C(S)^*)$ continuous$\}$. The letters $\mathcal{WLUC}$ stand for "weakly left uniformly continuous".

3.2. Theorem [Milnes (1973)]: Let $f \in C(S)$. The following are equivalent:

(a) $f \in \mathcal{WLUC}(S)$.

(b) $\text{coR}_S f$ is relatively compact in the topology p on $C(S)$ of pointwise convergence.

(c) $s \to \mu(L_s f)$ is continuous for all $\mu \in M(C(S))$.

Proof: (a) implies (b). Let $\{\sum_{s\in S} a_\alpha(s) R_s f\}$ be a net in $\text{coR}_S f$. For each t, α

(2) $\left(\sum_{s\in S} a_\alpha(s) R_s f\right)(t) = \sum_{s\in S} a_\alpha(s) f(ts) = \sum_{s\in S} a_\alpha(s) e(s)(L_t f)$,

where $e: S \to \beta S$ is the evaluation mapping. Now, there exists a subnet $\{\sum_{s\in S} a_\beta(s) e(s)\}$ which $\sigma(C(S)^*, C(S))$ converges to some $\mu \in C(S)^*$. Hence, from (2),

$$\lim_\beta \left(\sum_{s\in S} a_\beta(s) R_s f\right)(t) = \mu(L_t f),\ t \in S.$$

Since $t \to \mu(L_t f)$ is continuous, $\{\sum_{s \in S} a_\beta(s) R_s f\}$ p-converges in $C(S)$.

(b) implies (c). Let $\mu \in M(C(S))$. Then there exists a net $\{\sum_{s \in S} a_\alpha(s) e(s)\}$ which $\sigma(C(S)^*, C(S))$ converges to μ. By hypothesis, there exists a subnet $\{\sum_{s \in S} a_\beta(s) R_s f\}$ which p-converges in $C(S)$ to some function g. Thus, for each $t \in S$,

$$\mu(L_t f) = \lim_\beta \sum_{s \in S} a_\beta(s) e(s)(L_t f) = \lim_\beta \sum_{s \in S} a_\beta(s) R_s f(t)$$
$$= g(t),$$

hence $t \to \mu(L_t f)$ is continuous.

That (c) implies (a) follows from Proposition I.3.3.

3.3. <u>Corollary</u>: $\mathcal{WLUC}(S)$ is a translation invariant, left introverted, conjugate closed, norm closed, linear subspace of $C(S)$ containing the constant functions. Thus S has a $\mathcal{WLUC}$-affine compactification.

Proof: This follows from Lemma I.4.17 and Theorem 3.2.

3.4. <u>Theorem</u> [Rao (1965)]: $\mathcal{WLUC}(S)$ is the largest left introverted subspace of $C(S)$.

Proof: If F is a left introverted subspace of $C(S)$, then, for such $f \in F$ and $\mu \in C(S)^*$, the function $s \to \mu(L_s f)$ is in F (Remark I.4.12 (b)). In particular, $s \to \mu(L_s f)$ is continuous, hence $f \in \mathcal{WLUC}(S)$.

3.5. <u>Theorem</u>: Let X be a compact affine right topological semigroup and let $\psi: S \to X$ be a continuous homomorphism. Then (ψ, X) is a $\mathcal{WLUC}$-affine compactification of S if and only if it is maximal w.r.t. the following properties:

(a) $\overline{co}\psi(S) = X$.

(b) $\lambda_{\psi(s)}: X \to X$ is continuous for all $s \in S$.

Proof: This follows immediately from Theorem 3.2 and Corollary 1.7.

4. THE *LMC*-COMPACTIFICATION

4.1. Definition: $LMC(S) = LMC = \{f \in C(S) \mid s \to \mu(L_s f)$ is continuous for all $\mu \in \beta S\}$. *LMC* stands for "left multiplicatively continuous".

4.2. Theorem [Milnes (1973)]: Let $f \in C(S)$. The following are equivalent:

(a) $f \in LMC(S)$.

(b) $R_S f$ is relatively compact in the topology p on $C(S)$ of pointwise convergence.

Proof: Analogous to that of Theorem 3.2.

4.3. Corollary: $LMC(S)$ is a translation invariant left m-introverted C*-subalgebra of $C(S)$ containing the constant functions. Thus S has an *LMC*-compactification.

Proof: $LMC(S)$ is clearly closed under multiplication. Therefore, the corollary follows from Theorem 4.2 and Lemma I.4.18.

4.4. Theorem [Rao (1965)]: $LMC(S)$ is the largest left m-introverted subalgebra of $C(S)$.

Proof: Similar to that of 3.4.

4.5. <u>Theorem</u>: Let X be a compact right topological semigroup and let ψ: S → X be a continuous homomorphism. Then (ψ, X) is an *LMC*-compactification of S if and only if it is maximal w.r.t. the following properties:

(a) $\overline{\psi(S)} = X$.

(b) $\lambda_{\psi(s)}: X \to X$ is continuous for all $s \in S$.

Proof: Analogous to that of Theorem 3.5.

5. THE *LUC*-COMPACTIFICATION

5.1. <u>Definition</u>: $LUC(S) = LUC = \{f \in C(S) \mid s \to L_s f$ is norm continuous$\}$.

5.2. <u>Remarks</u>: The letters *LUC* stand for "left uniformly continuous". If S is a topological group then $f \in LUC(S)$ if and only if f is uniformly continuous with respect to the (right) uniformity generated by entourages of the form $\{(s,t) \mid st^{-1} \in V\}$, where V is a neighbourhood of the identity of S.

The proofs of the next two results are straightforward and are omitted.

5.3. <u>Theorem</u>: Let $f \in C(S)$. The following are equivalent:

(a) $f \in LUC(S)$.

(b) $R_S f$ is equicontinuous on S.

(c) $\mathrm{co}R_S f$ is equicontinuous on S.

5.4. <u>Lemma</u>: *LUC*(S) is a translation invariant left introverted C*-subalgebra of *C*(S) containing the constant functions. Thus S has an *LUC*-compactification.

5.5. <u>Theorem</u>: Let X be a compact right topological semigroup and let $\psi: S \to X$ be a continuous homomorphism. Then (ψ, X) is an LUC-compactification of S if and only if it is maximal w.r.t. the following properties:

(a) $\overline{\psi(S)} = X$.

(b) $(s,x) \to \psi(s)x: S \times X \to X$ is (jointly) continuous.

Proof: By Theorem 2.4 it suffices to show that if $\overline{\psi(S)} = X$, then the continuity of $(s,x) \to \psi(s)x$ is equivalent to $\psi^*C(X) \subset F$ and the continuity of the maps $\lambda_{\psi(s)}: X \to X$, $s \in S$, where $F = LUC(S)$.

Assume $(s,x) \to \psi(s)x$ is continuous, and let $h \in C(X)$. If $s, s_0 \in S$, then, since $\overline{\psi(S)} = X$,

$$||L_s\psi^*h - L_{s_0}\psi^*h|| = \sup_{x \in X} |h(\psi(s)x) - h(\psi(s_0)x)|.$$

Therefore, if $\{s_\alpha\}$ is a net in S converging to s_0,

$$||L_{s_\alpha}\psi^*h - L_{s_0}\psi^*h|| \to 0$$

by Lemma I.1.8 (a), hence $\psi^*h \in F$.

Conversely, suppose $\psi^*C(X) \subset F$ and $\lambda_{\psi(s)}: X \to X$ is continuous for each $s \in S$. We shall show that $(s,x) \to h(\psi(s)x)$ is continuous for any $h \in C(X)$. By Lemma I.1.8 (a) and the fact that $\overline{\psi(S)} = X$, it suffices to show that

$$s \to h(\psi(s)\psi(\cdot)): S \to C(S)$$

is norm continuous or, equivalently, that the family of mappings

$$s \to h(\psi(s)\psi(t)),\ t \in S,$$

is equicontinuous. But this family is precisely $R_S\psi^*h$ and, since $\psi^*h \in F$, the desired result follows from Theorem 5.3.

5.6. Remark: Lemma 5.4 implies that S also has an *LUC*-affine compactification. We omit the obvious affine analog of Theorem 5.5.

6. THE *K*-COMPACTIFICATION

6.1. Definition: $K(S) = K = \{f \in C(S) \mid R_S f$ is relatively compact-open compact$\}$. Recall that the compact-open topology on $C(S)$ is the (locally convex) topology of uniform convergence on compact subsets of S.

6.2. Lemma: $K(S)$ is a translation invariant left m-introverted C*-subalgebra of $C(S)$ containing the constant functions. Thus S has a *K*-compactification.

Proof: The mappings $f \to f^*$ and L_t, $t \in S$, are clearly compact-open continuous on $C(S)$, and an easy argument shows that $K(S)$ is closed under multiplication. Hence the lemma follows from Lemma I.4.18.

6.3. Theorem: Let X be a compact right topological semigroup and let ψ: $S \to X$ be a continuous homomorphism. Then (ψ, X) is a *K*-compactification of S if and only if it is maximal w.r.t. the following properties:

(a) $\overline{\psi(S)} = X$.

(b) $(s,x) \to \psi(s)x$: $K \times X \to X$ is (jointly) continuous for each compact subset $K \subset S$.

Proof: In view of 2.7 it is enough to show that if $\overline{\psi(S)} = X$ and if $\lambda_{\psi(s)}$: $X \to X$ is continuous for all $s \in S$, then property (b) is equivalent to the compact-open continuity of

$$x \to T_x h \colon X \to C(S),\ h \in C(X).$$

But this follows immediately from Lemma I.1.8 (a) and the fact that the uniformity on X is given by $C(X)$.

7. THE CK-AFFINE COMPACTIFICATION

7.1. Definition: $CK(S) = CK = \{f \in C(S) \mid \mathrm{co}R_S f$ is relatively compact-open compact$\}$.

The following two results have proofs analogous to those of the corresponding results in §6.

7.2. Lemma: $CK(S)$ is a translation invariant, left introverted, conjugate closed, norm closed, linear subspace of $C(S)$ containing the constant functions. Thus S has a CK-affine compactification.

7.3. Theorem: Let X be a compact right topological affine semigroup and let $\psi \colon S \to X$ be a continuous homomorphism. Then (ψ, X) is a CK-affine compactification of S if and only if it is maximal w.r.t. the following properties:

(a) $\overline{\mathrm{co}}\psi(S) = X$.

(b) $(s,x) \to \psi(s)x \colon K \times X \to X$ is (jointly) continuous for all compact sets $K \subset S$.

8. THE WAP-COMPACTIFICATION

8.1. Definition: $WAP(S) = WAP = \{f \in C(S) \mid R_S f$ is relatively weakly compact$\}$. The letters WAP stand for "weakly almost periodic"; "weakly compact" means "$\sigma(C(S), C(S)^*)$ compact".

8.2. <u>Theorem</u>: Let $f \in C(S)$. The following are equivalent:

(a) $f \in \mathcal{WAP}(S)$.

(b) $\text{co}R_S f$ is relatively weakly compact.

(c) $\lim_m \lim_n f(s_m t_n) = \lim_n \lim_m f(s_m t_n)$ whenever $\{s_m\}$ and $\{t_n\}$ are sequences in S such that all of the limits exist.

(d) $L_S f$ is relatively weakly compact.

(e) $\text{co}L_S f$ is relatively weakly compact.

(f) $R_S f$ is relatively $\sigma(C(S),\beta S)$ compact.

(g) $L_S f$ is relatively $\sigma(C(S),\beta S)$ compact.

Proof: The equivalence of (a) and (b) and of (d) and (e) follows from the theorem of Krein-Smulian [Dunford and Schwartz (1964); p. 434].

(a) implies (c). Let $e: S \to \beta S$ denote the evaluation map, and let $\{s_m\}$ and $\{t_n\}$ be sequences in S such that all of the limits in (c) exist. Then there exist subnets $\{s_{m_\alpha}\}$ and $\{t_{n_\beta}\}$, and $\mu \in \beta S$, $g \in C(S)$, such that $\{e(s_{m_\alpha})\}$ $\sigma(C(S)^*,C(S))$ converges to μ and $\{R_{t_{n_\beta}} f\}$ $\sigma(C(S),C(S)^*)$ converges to g. Therefore

$$\lim_m \lim_n f(s_m t_n) = \lim_m \lim_n e(s_m)(R_{t_n} f)$$

$$= \lim_m e(s_m)(g) = \mu(g),$$

and

$$\lim_n \lim_m f(s_m t_n) = \lim_n \lim_m e(s_m)(R_{t_n} f)$$

$$= \lim_n \mu(R_{t_n} f) = \mu(g).$$

(d) implies (c). Analogous to the proof that (a) implies (c).

(c) implies (g). Let $e^*: C(\beta S) \to C(S)$ denote the adjoint of e. We must show that $G = e^{*-1}(L_S f)$ is relatively compact in the topology of pointwise convergence on $C(\beta S)$, i.e., that the pointwise closure $\overline{G}$ of G in $B(\beta S)$ contains only continuous functions. A simple argument shows that a function $g \in B(\beta S)$ is continuous if and only if for each $x \in \beta S$ and each net $\{e(s_\alpha)\}$ converging in βS to x, $\lim_\alpha g(e(s_\alpha)) = g(x)$. Hence, if $g \in \overline{G}$ is not continuous, then there exist $x \in \beta S$ and $\varepsilon > 0$ such that every neighbourhood V in βS of x contains a point e(t) for which $|g(x) - g(e(t))| \geq \varepsilon$. Choose any function $g_1 \in G$, and let $t_1 \in S$ be such that

$$|g_1(x) - g_1(e(t_1))| < 1, \quad |g(x) - g(e(t_1))| \geq \varepsilon.$$

Choose $g_2 \in G$ such that

$$|g_2(e(t_1)) - g(e(t_1))| < 1, \quad |g_2(x) - g(x)| < 1.$$

Let $t_2 \in S$ be such that, for i = 1, 2,

$$|g_i(e(t_2)) - g_i(x)| < \tfrac{1}{2}, \quad |g(x) - g(e(t_2))| \geq \varepsilon.$$

Choose $g_3 \in G$ such that for i = 1, 2,

$$|g_3(e(t_i)) - g(e(t_i))| < \tfrac{1}{2}, \quad |g_3(x) - g(x)| < \tfrac{1}{2}.$$

Continuing in this manner we obtain sequences $\{g_m\}$ in G and $\{t_n\}$ in S such that, for each n,

$$|g_i(e(t_n)) - g_i(x)| < \tfrac{1}{n}, \quad i = 1, 2, \ldots, n,$$

$$|g_{n+1}(e(t_i)) - g(e(t_i))| < \tfrac{1}{n}, \quad i = 1, 2, \ldots, n,$$

$$|g_{n+1}(x) - g(x)| < \tfrac{1}{n},$$

$$|g(x) - g(e(t_n))| \geq \varepsilon.$$

For each m there exists $s_m \in S$ such that $e^*g_m = L_{s_m} f$. Then

$$\lim_n \lim_m f(s_m t_n) = \lim_n \lim_m g_m(e(t_n)) = \lim_n g(e(t_n)),$$

$$\lim_m \lim_n f(s_m t_n) = \lim_m \lim_n g_m(e(t_n)) = \lim_m g_m(x) = g(x)$$

and, taking a subsequence of $\{t_n\}$ if necessary, we see that (c) cannot hold.

(g) implies (d). If $L_S f$ is $\sigma(C(S), \beta S)$ compact, then $e^{*-1}(L_S f)$ is relatively compact in the weak topology of $C(\beta S)$ [Grothendieck (1952)]. Therefore $L_S f$ is $\sigma(C(S), C(S)^*)$ compact.

(c) implies (f) and (f) implies (a). Similar to the proofs that (c) implies (g) and (g) implies (d).

8.3. <u>Lemma</u>: $\mathcal{WAP}(S)$ is a translation invariant, left and right introverted, C*-subalgebra of $C(S)$ containing the constant functions. Hence S has a $\mathcal{WAP}$-compactification.

Proof: The mappings $f \to f^*$ and L_t on $C(S)$ are clearly $\sigma(C(S), C(S)^*)$ continuous. Hence from 8.2 (b) and from Lemma I.4.17 it follows that $\mathcal{WAP}$ is a translation invariant, left introverted, conjugate closed, norm closed, linear subspace of $C(S)$ containing the constant functions. Similarly, using 8.2 (e), one shows that $\mathcal{WAP}$ is right introverted.

It remains to prove that $\mathcal{WAP}$ is closed under (pointwise) multiplication. Let f, $g \in \mathcal{WAP}$ and let $\{s_n\}$ be any sequence in S. By the Eberlein-Smulian Theorem [Dunford and Schwartz (1964); p. 430], there exist a subsequence $\{t_n\}$ and f_0, $g_0 \in C(S)$ such that $\{f_n\} = \{R_{t_n} f\}$ and $\{g_n\} = \{R_{t_n} g\}$ $\sigma(C(S), C(S)^*)$

converge to f_0 and g_0 respectively. We shall show that $\{f_n g_n\}$ $\sigma(C(S),C(S)^*)$ converges to $f_0 g_0$. For this we may assume S is compact; otherwise, replace $C(S)$ by $C(\beta S)$. Then, since $f_n g_n \to f_0 g_0$ pointwise, Grothendieck's Theorem assures that $\{f_n g_n\}$ converges in the $\sigma(C(S),C(S)^*)$ topology. Therefore $fg \in WAP(S)$.

8.4. Theorem: Let X be a compact right topological semigroup and let $\psi: S \to X$ be a continuous homomorphism. Then (ψ,X) is a WAP-compactification if and only if it is maximal w.r.t. the following properties:

(a) $\overline{\psi(S)} = X$.

(b) X is a semitopological semigroup.

Proof: By 2.7 it suffices to prove that if $\overline{\psi(S)} = X$, then X is semitopological if and only if

$$x \to T_x h: X \to C(S)$$

is $\sigma(C(S),C(S)^*)$ continuous for all $h \in C(X)$. Note first that, if $h \in C(X)$, then $T_x h = \psi^*(R_x h)$. Furthermore, if $\overline{\psi(S)} = X$, then ψ^* is an isometry and $R_x h = \psi^{*-1}(T_x h)$. Hence the result follows from Corollary I.1.9.

8.5. Corollary: Let F be a left m-introverted translation invariant C*-subalgebra of $C(S)$ containing the constant functions and suppose (ϕ,Y) is an F-compactification of S. Then Y is a semitopological semigroup if and only if $F \subset WAP(S)$.

Proof: Let (ψ,X) be a WAP-compactification of S. If Y is semitopological, then by 8.4 there exists a continuous homomorphism $\theta: X \to Y$ such that $\theta \circ \psi = \phi$. Since

$\phi^*(C(Y)) = F$ and $\psi^*(C(X)) = WAP$ we have

$$F = \psi^* \circ \theta^*(C(Y)) \subset WAP.$$

Conversely, if $F = \phi^*(C(Y)) \subset WAP$, then by 2.4 there exists a continuous homomorphism $\theta: X \to Y$ such that $\theta \circ \psi = \phi$. Since θ is surjective, Y is semitopological.

8.6. Remarks: (a) Lemma 8.3 implies that S has also a WAP-affine compactification. We leave it to the reader to formulate the affine analogs of 8.4 and 8.5.

(b) Let (ψ, X) be the canonical WAP-compactification of S. If $L(WAP)$, the space of bounded linear operators on WAP, is given the weak operator topology, then $L(WAP)$ is a semitopological semigroup (under composition), and the mapping $\nu \to T_\nu: X \to L(WAP)$, defined in I.4.9, is a continuous homomorphism. Furthermore, if S has a left identity then this mapping is injective. Since $T_{\psi(s)} = R_s$, it follows that (R, T_X) is a WAP-compactification of S. This is the WAP-compactification discussed in [deLeeuw and Glicksberg (1961)] and [Burckel (1970)]. Note that T_X is the weak operator closure of R_S.

The following theorem is an application of previous results to the structure theory of compact affine right topological semigroups.

8.7. Theorem: Let T be a compact affine right topological semigroup with a compact semitopological subsemigroup S such that $\overline{\text{co}S} = T$. If $\lambda_s: T \to T$ is continuous for each $s \in S$, then T is semitopological.

Proof: Since S is compact, Corollary I.1.9 implies that

$$WAP(S) = C(S) = WLUC(S).$$

Therefore (e,M(C(S))) is both a WLUC-affine compactification of S and a WAP-affine compactification of S. By the maximal property of WLUC-compactifications, there exists a continuous surjective homomorphism ϕ: M(C(S)) → T. Since M(C(S)) is semitopological, it follows that T is semitopological.

Our next result will be useful in later sections.

8.8. Lemma: Let F be a translation invariant, conjugate closed, norm closed, linear subspace of C(S) containing the constant functions. If F ⊂ WAP(S), then F is left and right introverted.

Proof: We prove only left introversion; the proof of right introversion is similar. Let f ∈ F. If μ is a finite mean on C(S), then $T_\mu f \in \text{co}R_S f \subset F$. Since f ∈ WAP, the mapping $\mu \to T_\mu f$: M(C(S)) → WAP is $\sigma(C(S), C(S)^*)$ continuous. Therefore, if $\{\mu_\alpha\}$ is a net of finite means on C(S) converging in M(C(S)) to μ, then $\{T_{\mu_\alpha} f\}$ is a net in F converging weakly to $T_\mu f$. Since F is weakly closed, $T_\mu f \in F$.

8.9. Remarks: (i) We note here that Theorem II.5.6 and Corollary II.5.8 show the connection between the existence of a (left) invariant mean on WAP(S) and the structure of the minimal ideal K(X), where X is a WAP-compactification of S. In particular, WAP(S) has an invariant mean if and only if K(X) is a compact topological group. When this is the case, it follows that SAP(S) = AP(S) (of §§ 9 and 10 ahead) and K(X) is an AP-compactification of S. See [deLeeuw and Glicksberg (1961)] and Theorem 10.6 ahead.

(ii) We make a connection with the mean ergodic theorem. Let ϕ be a representation of a group G by uniformly bounded operators on an L^p space B, $1 < p < \infty$. Then $\overline{co}(\phi(G)b)$ is weakly compact, $b \in B$, and $T = \overline{co}\phi(G)$ (strong (or weak) operator closure) is a compact affine semitopological semigroup in the weak operator topology. It follows from the affine analog of 8.4 and IV.1.14 and II.5.7, that K(T) consists of a single member z, a zero for T. Hence, for all $b \in B$, zb is the only invariant element in $\overline{co}(\phi(G)b)$, and there is a net $\{A_\alpha\} \in co\phi(G)$ such that $||A_\alpha b - zb|| \to 0$, $b \in B$. See [Greenleaf (1973), Milnes (1977)] for details, variants and generalizations.

9. THE AP-COMPACTIFICATION

9.1. Definition: $AP(S) = AP = \{f \in C(S) \mid R_S f$ is relatively norm compact$\}$. The letters AP stand for "almost periodic".

9.2. Theorem: Let $f \in C(S)$. The following are equivalent:

(a) $f \in AP(S)$.

(b) $L_S f$ is relatively norm compact.

(c) $coR_S f$ is relatively norm compact.

(d) $coL_S f$ is relatively norm compact.

Proof: The equivalence of (a) and (c) and of (b) and (d) follow from Mazur's Theorem [Dunford and Schwartz (1964); p. 416].

(a) implies (b). We anticipate the result, proved below, that S has an AP-compactification (ψ, X), where X is a compact topological semigroup. Let $f \in AP(S)$. Then, since $AP(S) = \psi^*(C(X))$, there exists a function $g \in C(X)$ such that $f = \psi^*(g)$. Thus

$$L_s f = \psi^*(L_{\psi(s)} g) \in \psi^*(L_X g),$$

and by Corollary I.1.9, $L_S f$ is relatively norm compact.

(b) implies (a). Let T denote the semigroup S with multiplication $s * t = ts$ (where ts is the original product in S). Then the result follows from the argument used to prove that (a) implies (b).

9.3. Lemma: $AP(S)$ is a translation invariant, left and right introverted, C*-subalgebra of $C(S)$ containing the constant functions. Hence S has an AP-compactification.

Proof: Similar to (but easier than) that of 8.3.

9.4. Theorem: Let X be a compact right topological semigroup and let $\psi: S \to X$ be a continuous homomorphism. Then (ψ, X) is an AP-compactification if and only if it is maximal w.r.t. the following properties:

(a) $\overline{\psi(S)} = X$.

(b) X is a topological semigroup.

Proof: Similar to the proof of 8.4.

9.5. Corollary: Let F be a translation invariant left m-introverted C*-subalgebra of $C(S)$ containing the constant functions, and let (ϕ, Y) be an F-compactification of S. Then Y is a topological semigroup if and only if $F \subset AP(S)$.

Proof: Similar to that of 8.5.

9.6. <u>Remarks:</u> (a) Lemma 9.3 implies that S has also an $\mathcal{AP}$-affine compactification. The reader may easily formulate the affine analogs of 9.4 and 9.5.

(b) If S has a left identity and if (ψ, X) is the canonical $\mathcal{AP}$-compactification, then the mapping

$$\nu \to T_\nu : X \to L(\mathcal{AP})$$

is a continuous injective homomorphism, when $L(\mathcal{AP})$ is given the strong operator topology. Since $T_{\psi(s)} = R_s$, it follows that (R, T_X) is an $\mathcal{AP}$-compactification of S, and T_X is the strong operator closure of R_S. (R, T_X) is the $\mathcal{AP}$-compactification discussed in [deLeeuw and Glicksberg (1961)] and [Burckel (1970)].

The proof of the next result is similar to that of 8.7.

9.7. <u>Theorem:</u> Let T be a compact affine right topological semigroup with a compact topological subsemigroup S such that $\overline{\text{co}}S = T$. If $\lambda_s : T \to T$ is continuous for each $s \in S$, then T is topological.

9.8. <u>Remark:</u> It follows from the fact that a compact topological semigroup with a dense subgroup is also a group that $\mathcal{AP}$-compactifications of groups are (compact topological) groups.

10. THE *SAP*-COMPACTIFICATION

10.1. Definition: If H is a Hilbert space and $L(H)$ is the algebra of all bounded, linear operators on H, then a weakly continuous unitary representation of a semitopological semigroup S is a homomorphism U of S into the group of unitary operators in $L(H)$ such that the functions

(3) $$s \to (U_s\xi,\eta), \quad \xi, \eta \in H,$$

are continuous. The functions (3) are called the coefficients of the representation U, and U is called finite dimensional if H is finite dimensional.

10.2. Definition: $SAP(S) = SAP$ is the norm closure in $C(S)$ of the linear span of the coefficients of finite dimensional weakly continuous unitary representations of S. The letters *SAP* stand for "strongly almost periodic".

10.3. Lemma: $SAP(S)$ is a translation invariant left and right introverted C*-subalgebra of $C(S)$ containing the constant functions. Hence S has an *SAP*-compactification.

Proof: By the device of tensor products, one shows that the product of two coefficients of finite dimensional unitary representations is itself such a function, and from this it easily follows that *SAP* is a C*-algebra. Clearly *SAP* contains 1.

To show that *SAP* is translation invariant simply note that if $f(s) = (U_s\xi,\eta)$ then

$$(L_tf)(s) = (U_s\xi,\omega) \text{ and } (R_tf)(s) = (U_s\zeta,\eta),$$

where $\omega = U_t^*\eta$ and $\zeta = U_t\xi$.

The left and right introversion of SAP will follow from 8.8 if we show that $SAP \subset WAP$. We shall show, in fact, that $SAP \subset AP$. Let $f(s) = (U_s\xi,\eta)$, where

$$s \to U_s\colon S \to L(H)$$

is a finite dimensional unitary representation of S. Define a bounded linear operator $Q\colon H \to SAP$ by

$$(Q\zeta)(s) = (U_s\zeta,\eta),\ \zeta \in H.$$

Then $QU_t\xi = R_tf$, hence $R_Sf = QU_S\xi$. Since $U_S\xi$ is a bounded subset of a finite dimensional space, R_Sf must be relatively norm compact. Therefore $SAP \subset AP$.

10.4. Theorem: Let X be a compact right topological semigroup and let $\psi\colon S \to X$ be a continuous homomorphism. Then (ψ,X) is an SAP-compactification if and only if it is maximal w.r.t. the following properties:

(a) $\overline{\psi(S)} = X$.

(b) X is a topological group.

Proof: By 2.4 it suffices to show that if $\overline{\psi(S)} = X$ and $\lambda_{\psi(s)}\colon X \to X$ is continuous for all $s \in S$, then X is a topological group if and only if $\psi^*C(X) \subset SAP(S)$.

Now, if X is a compact topological group, then by the Gelfand-Raikov Theorem [Hewitt and Ross (1963); p. 343], $SAP(X)$ separates the points of X. Therefore $SAP(X) = C(X)$ (Stone-Weierstrass Theorem), hence $\psi^*C(X) \subset SAP(S)$.

Conversely, suppose $\psi^*C(X) \subset SAP(S)$. Then if (ϕ,Y) is any SAP-compactification of S, there exists a continuous surjective homomorphism $\theta\colon Y \to X$ (Theorem 2.4). Hence X will be seen to be a topological group provided we show that Y is a topological group. Since $SAP(S) \subset AP(S)$ (proof of

10.3), Y is a compact topological semigroup (9.5). Hence it suffices to show that Y is cancellative [Hofmann and Mostert (1966); p. 77, Ex. 9].

Let x, y, z $\in$ Y with xz = yz. To show that x = y it suffices to show that $\hat{f}(x) = \hat{f}(y)$ for all f $\in$ $SAP(S) = \phi^*C(Y)$, where $\phi^*\hat{f} = f$. For this we may take f to be of the form $f(s) = (U_s\xi,\eta)$, where

$$s \to U_s\colon S \to L(H)$$

is a weakly continuous finite dimensional unitary representation on the Hilbert space H. Let $\{s_\alpha\}$ be a net in S such that $\{\phi(s_\alpha)\}$ converges to z. There exist a subnet $\{s_\beta\}$ and a unitary operator U $\in$ $L(H)$ such that $\{U_{s_\beta}\}$ converges to U in $L(H)$. Let g $\in$ $SAP(S)$ be the function $g(s) = (U_sU^{-1}\xi,\eta)$. Then, for any s $\in$ S,

$$(R_z\hat{g})(\phi(s)) = \hat{g}(\phi(s)z) = \lim_\beta g(ss_\beta) = f(s),$$

so $\hat{f} = R_z\hat{g}$ and thus

$$\hat{f}(x) = \hat{g}(xz) = \hat{g}(yz) = \hat{f}(y).$$

The proof that Y is left cancellative is similar.

10.5. <u>Corollary</u>: Let F be a translation invariant left m-introverted C*-subalgebra of $C(S)$ containing the constant functions, and let (ϕ,Y) be an F-compactification of S. Then Y is a topological group if and only if $F \subset SAP(S)$.

10.6. <u>Theorem</u>: Let F be a translation invariant left m-introverted C*-subalgebra of $C(S)$ containing the identity and suppose $F \subset WAP(S)$. Let (ψ,X) be an F-compactification of S and e $\in$ E(K(X)). Then F is left amenable if and only if T_e $(= \psi^*R_e\psi^{*-1})$ is a retraction of F onto $F \cap SAP(S)$.

In this case, $(\rho_e\psi, Xe)$ is an $F \cap SAP(S)$-compactification of S.

Proof: Assume first that F is left amenable. Then Xe is a compact topological group (II.5.6) and $\rho_e\psi$: $S \to Xe$ a continuous homomorphism with range dense in Xe. By Theorem 10.4, then, there exists a continuous homomorphism ϕ: $MS \to Xe$ such that $\phi m = \rho_e\psi$, where (m,MS) denotes the canonical SAP-compactification of S. Hence we have

$$T_eF = \psi^*R_eC(X) = \psi^*\rho_e^*C(Xe) = m^*\phi^*C(Xe) \subset SAP(S).$$

Now let (γ,Y) be a $F \cap SAP(S)$ compactification of S. Then Y is a compact topological group, and there exists a continuous homomorphism θ: $X \to Y$ such that $\theta\psi = \gamma$. If $f \in F \cap SAP(S)$, then since $\theta(e)$ is the identity of Y,

$$T_ef = \psi^*R_e\theta^*\gamma^{*-1}f = \psi^*\theta^*\gamma^{*-1}f = f.$$

Therefore T_e is a retraction onto $F \cap SAP(S)$. In particular,

$$(\rho_e\psi)^*C(Xe) = T_eF = F \cap SAP(S),$$

hence $(\rho_e\psi, Xe)$ is an $F \cap SAP$-compactification of S.

Conversely, assume T_e is a retraction of F onto $F \cap SAP(S)$. Let μ be the invariant mean on $SAP(S)$. Then if $s \in S$ and $f \in F$ we have $(T_e^*\mu)(L_sf) = \mu(\psi^*R_eL_{\psi(s)}\psi^{*-1}f) = \mu(\psi^*L_{\psi(s)}R_e\psi^{*-1}f) = \mu(L_sT_ef) = \mu(T_ef) = (T_e^*\mu)(f)$, so $T_e^*\mu \in LIM(F)$.

11. THE LWP-COMPACTIFICATION

11.1. Definition: $LWP(S) = LWP = WAP(S) \cap LUC(S)$.

11.2. Lemma: $LWP(S)$ is a translation invariant left and right introverted C*-subalgebra of $C(S)$ containing the constant functions. Therefore S has an LWP-compactification.

Proof: The result follows from Lemmas 5.4, 8.3, and 8.8.

11.3. Theorem: Let X be a compact right topological semigroup and let $\psi: S \to X$ be a continuous homomorphism. Then (ψ, X) is an LWP-compactification of S if and only if it is maximal w.r.t. the following properties:

(a) $\overline{\psi(S)} = X$.

(b) X is a semitopological semigroup.

(c) $(s,x) \to \psi(s)x: S \times X \to X$ is continuous.

Proof: Combine the proofs of 5.5 and 8.4.

11.4. Remark: By 11.2, S also has an LWP-affine compactification. The reader may easily formulate the affine analog of 11.3.

12. THE KWP-COMPACTIFICATION

12.1. Definition: $KWP(S) = KWP = K(S) \cap WAP(S)$.

12.2. Lemma: $KWP(S)$ is a translation invariant left and right introverted C*-subalgebra of $C(S)$ containing the constant functions. Thus S has a KWP-compactification.

Proof: This follows from Lemmas 6.2, 8.3, and 8.8.

12.3. Theorem: Let X be a compact right topological semigroup and let ψ: S $\to$ X be a continuous homomorphism. Then (ψ, X) is a *KWP*-compactification if and only if it is maximal w.r.t. the following properties:

(a) $\overline{\psi(S)} = X$.

(b) X is a semitopological semigroup.

(c) $(s,x) \to \psi(s)x: K \times X \to X$ is continuous for all compact subsets $K \subset S$.

Proof: Combine the proofs of 6.3 and 8.4.

13. THE *CKWP*-AFFINE COMPACTIFICATION

13.1. Definition: $CKWP(S) = CKWP = CK(S) \cap WAP(S)$.

The following results have proofs analogous to the proofs of the corresponding results in §12.

13.2. Lemma: *CKWP*(S) is a translation invariant left and right introverted, conjugate closed, norm closed, linear subspace of *C*(S) containing the constant functions. Thus S has a *CKWP*-affine compactification.

13.3. Theorem: Let X be a compact right topological affine semigroup and let ψ: S $\to$ X be a continuous homomorphism. Then (ψ, X) is a *CKWP*-affine compactification if and only if it is maximal w.r.t. the following properties:

(a) $\overline{\text{co}}\psi(S) = X$.

(b) X is a semitopological semigroup.

(c) $(s,x) \to \psi(s)x: K \times X \to X$ is continuous for all compact subsets $K \subset S$.

14. INCLUSION RELATIONSHIPS AMONG THE SUBSPACES

In this section we consider what inclusion relationships hold among the subspaces of this chapter; we will consider inclusions that always hold and inclusions that hold if (and sometimes only if) the semitopological semigroup in question is required to have some extra algebraic or topological properties. For the convenience of the reader, we collect here in one place definitions of all spaces considered.

14.1. Definition: Let S be a semitopological semigroup. We collect here definitions from the previous eleven sections.

$WLUC = \{f \in C(S) \mid s \to L_s f \text{ is } \sigma(C(S),C(S)^*) \text{ continuous}\}$.

$LMC = \{f \in C(S) \mid s \to \mu(L_s f) \text{ is continuous for all } \mu \in \beta S\}$.

$LUC = \{f \in C(S) \mid s \to L_s f \text{ is norm continuous}\}$.

$K = \{f \in C(S) \mid R_S f \text{ is relatively compact-open compact}\}$.

$CK = \{f \in C(S) \mid coR_S f \text{ is relatively compact-open compact}\}$.

$WAP = \{f \in C(S) \mid R_S f \text{ is relatively weakly compact}\}$.

$AP = \{f \in C(S) \mid R_S f \text{ is relatively norm compact}\}$.

SAP is the norm closure in $C(S)$ of the linear span of the coefficients of finite dimensional weakly continuous unitary representations of S. And, finally,

$LWP = LUC \cap WAP$, $KWP = K \cap WAP$, $CKWP = CK \cap WAP$.

14.2. Notation: We establish here notation for the canonical (affine) compactifications associated with the spaces of 14.1 (recalling that the symbol e has been used to denote the canonical map of S into $\beta S \subset C(S)^*$ or into F^*, where F is a general subspace of $C(S)$).

Subspace	Canonical affine compactification
WLUC	(cp,cPS)
CK	(ck,cKS)
CKWP	(cq,cQS)

Subspace	Canonical compactification
LMC	(p,PS)
K	(k,KS)
LUC	(u,US)
WAP	(w,WS)
AP	(a,AS)
SAP	(m,MS)
KWP	(q,QS)
LWP	(v,VS)

When necessary, the notation will be made to reflect which semigroup is under consideration, e.g., the canonical *LMC*(S)-compactification will be denoted (p_S,PS). Also, the reader is reminded that, for example, $p(S) = \{p(s) \mid s \in S\}$ is dense in PS, while it is the convex hull (in cKS) of ck(S) that is dense in cKS. Of the last eight subspaces, *LUC*, *WAP*, *AP*, *SAP* and *LWP* are known to be left introverted and hence have canonical affine compactifications as well, notation for which will be, for example,

LUC (cu,cUS).

14.3. Inclusion Diagram: For any semitopological semigroup S, the following inclusions always hold.

(All arrows indicate inclusions.)

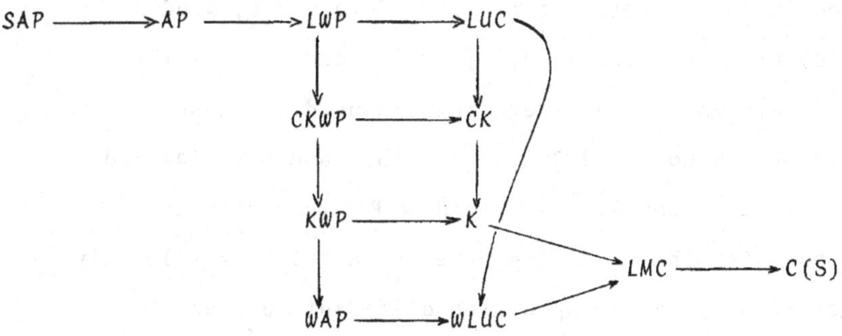

14.4. Remarks: (i) The inclusion $SAP \subset AP$ has been demonstrated (proof of Lemma 10.3), and $WAP \subset WLUC$ follows from Theorems 8.2 ((a) and (b)) and 3.2. That $LUC \subset CK$ follows from Theorem 5.3. The other inclusions of 14.3 then follow easily.

(ii) No example is known for which $LUC \neq CK$ (or $LWP \neq CKWP$), $CK \neq K$ (or $CKWP \neq KWP$) or $WLUC \neq LMC$. Examples are known (Examples V.2, 1 and 2) for which all the other inclusions of 14.3 are proper. No example is known for which $K \not\subset WLUC$, but $WLUC$ can properly contain K. And Theorem 8.2 ((a) and (b)), respectively 9.2 ((a) and (c)), is the reason the notation $CWAP$, respectively CAP, is superfluous. Also, since the canonical compactification associated with a subspace $CSAP$, if such were defined, should be a compact affine topological group, and, therefore, a trivial group (Corollary III.4.4), the natural definition of *CSAP* is the space of constant functions $\mathbb{C}$.

(iii) Many of the spaces considered here have been defined in terms of left translates (e.g., *LUC*, *WLUC* and *LMC*) or right orbits (e.g., *AP*, *WAP* and *K*). If the words "left" and "right" were interchanged in these definitions, Theorems 9.2 ((a) and (b)) and 8.2 ((a) and (d)) show the same spaces *AP* and *WAP*, respectively, would be determined. On the other hand, if S is a locally compact topological group G with different left and right uniformities, Remark 5.2 (and Theorem 14.6 ahead) shows that a different subspace of C(G) would be defined in place of *LUC* (= *K* = *WLUC* = *LMC* — the subspace would be called the right uniformly continuous subspace of C(S)).

14.5. Remarks: We state here some containments that hold true in special cases. Unless otherwise indicated, the containments can be proper.

(i) If S is discrete, then

$$SAP \subset AP \subset LWP = CKWP = KWP = WAP \subset LUC = CK = K = LMC = C(S)$$

(ii) If S is a compact semitopological semigroup, then

$$SAP \subset AP = LWP = LUC = CKWP = CK = KWP = K \subset WAP = WLUC = LMC = C(S).$$

(The containment $AP \subset LUC$ always holds, and the reverse containment is easily seen to hold in this setting.)

(iii) If S is a compact topological semigroup, then

$$SAP \subset AP = LWP = LUC = CKWP = CK = KWP = K = WAP = WLUC = LMC = C(S) .$$

(iv) If S is a compact topological group, then all twelve spaces are equal. (This follows from the theorem of Weyl and Peter, Corollary (22.14) of [Hewitt and Ross, (1963)]).

(v) If S is algebraically a group, then

$$SAP = AP$$

because a compact topological semigroup containing a dense subgroup must itself be a group (A.1.5 of [Hofmann and Mostert (1966)]).

A completeness assumption is one of the hypotheses of the following theorem (and it is necessary; see Example V.2.5). We do not define it explicitly, but remark that locally compact spaces and complete metric spaces are complete in this sense and refer the reader to [Namioka (1974)], where the theorem we quote is proved.

14.6. Theorem: Let G be a semitopological semigroup that is algebraically a group, and suppose that, as a topological space, G is strongly countably complete and regular. Then $LUC(G) = LMC(G)$.

Proof: By Theorem 4.5, the map $(s,\mu) \to p(s)\mu$ from $G \times PG$ into PG is separately continuous. (Notation is as in 14.3.) In the setting here, Theorem 3.1 of [Namioka (1974)] tells us this map is (jointly) continuous. If we then identify functions $f \in LMC$ with functions on $G \times PG$ (via the Gelfand transform, i.e.,

$$f: (s,\mu) \to \mu(L_s f) = p(s)\mu(f)),$$

Lemma I.1.8 (a) gives the desired conclusion.

14.7. <u>Definition</u>: A topological space Y is called a <u>k-space</u> (a <u>ck-space</u>) if it has the property that a subset A is closed whenever $A \cap K$ is closed in K for each compact (countably compact) $K \subset Y$.

First countable and locally compact spaces are k-spaces and k-spaces are ck-spaces. A function on a k-space (ck-space) Y is continuous if and only if its restriction to each compact (countably compact) subset of Y is continuous.

We remark that it is not known if the conclusion of the following theorem ever fails to hold.

14.8. <u>Theorem</u>: Let S be a semitopological semigroup and a k-space. Then $LUC = CK = K$.

Proof: K always contains LUC. The reverse inclusion follows from the mapping property of KS (Theorem 6.3), Lemma I.1.8 (a) and the property of k-spaces mentioned above (i.e., arguing as in the last sentence of Theorem 14.6, one shows that the function $s \to L_s f$: $S \to C(S)$ is continuous on each compact subset of S; hence, this function is continuous).

Some alternate characterizations of *WLUC* and *LMC* have been given in §§ 3, 4. We next show that these spaces are identical in all familiar cases.

14.9. <u>Theorem</u>: Let S be a semitopological semigroup and a ck-space. Then each of the following assertions about a function $f \in C(S)$ is equivalent to any of the others.

(i) $f \in LMC$.

(ii) For each countably compact $A \subset S$, $\{L_s f \mid s \in A\}$ is $\sigma(C(S),\beta S)$ compact.

(iii) For each countably compact $A \subset S$, $\{L_s f \mid s \in A\}$ is weakly compact.

(iv) $\lim_i \lim_j f(s_i t_j)$ equals $\lim_j \lim_i f(s_i t_j)$ whenever $\{t_j\}$ is a sequence in S and $\{s_i\}$ is a sequence contained in a countably compact subset of S such that all the limits exist.

(v) $f \in WLUC$.

Proof: For much of this proof, we identify $C(S)$ and $C(\beta S)$. We prove first that (i) implies (ii). Let $f \in LMC$ and let $A \subset S$ be countably compact. Since the function $s \to L_s f$ from S into $C(S)$ is $\sigma(C(S),\beta S)$ continuous, $\{L_s f \mid s \in A\}$ is countably $\sigma(C(S),\beta S)$ compact and hence $\sigma(C(S),\beta S)$ compact [Grothendieck (1952); Théorème 2].

That (ii), (iii) and (iv) are equivalent follows from [Grothendieck (1952); Théorèmes 5 and 6]. To prove that

(iii) implies (v), it suffices to show that, for any $\mu \in C(S)^*$ and countably compact $A \subset S$, the restriction of the function $s \to \mu(L_s f)$ to A is continuous on A. However, this follows directly from (iii). And $WLUC \subset LMC$ always.

14.10. Remarks: (a) Worthy of special note is the striking similarity of the Theorem 14.9 (iv) characterization of LMC (in the ck-space setting) to the Theorem 8.2 (c) characterization of WAP.

(b) Some topological properties for S well-suited to the present line of proof that $LMC = WLUC$ are that every convergent net in S should have a subnet which is contained in a countably compact subset of S (which is the case if S is first countable or locally compact), or that a (bounded) function continuous on every countably compact subset of S should be continuous on S. It has been remarked that ck-spaces have the latter property.

(c) For a function $f \in C(S)$, conditions (ii), (iii) and (iv) are equivalent without the ck-space hypothesis.

(d) We remind the reader that it is not known if the containment $WLUC \subset LMC$ is ever proper.

14.11. Inclusion Diagram for Many Groups: Summarizing the last few results we see that the inclusion diagram 14.3 becomes much simpler for members of a broad class of topological groups G, including those that are locally compact or complete metric. We have

$$SAP = AP \subset LWP = CKWP = KWP = WAP$$
$$\subset LUC = CK = K = WLUC = LMC \subset C(G).$$

In case G' is a dense subgroup of the kind of group considered here, we still have $\mathcal{WAP}(G') \subset \mathcal{LUC}(G')$ (Corollary 15.7 ahead), but the equality $\mathcal{LUC}(G') = \mathcal{LMC}(G')$ can be lost (Example V.2.5).

14.12. Theorem: (a) The following conditions on a topological group G are equivalent.

(i) $\mathcal{AP} = C(G)$.

(ii) $\mathcal{WAP} = C(G)$.

(iii) G is pseudocompact.

(b) If G is a totally bounded topological group, then $\mathcal{AP} = \mathcal{WAP}$.

Proof: (b) follows from Corollary 15.7 ahead, Remark 14.5 (iv) and a classic theorem on the completion of topological groups [Weil (1937); p. 32]. We turn to (a). That (i) is equivalent to (iii) is proved in [Comfort and Ross (1966)]; that (i) implies (ii) is trivial. Thus we will be done when we prove (ii) implies (i) and we achieve this by proving that the assumption $\mathcal{WAP} = C(G)$ implies G is totally bounded; (b) then completes the job.

Now, if G is not totally bounded, there is a neighbourhood V of $e \in G$ and a sequence $\{s_n\} \subset G$ such that

$$* \qquad Vs_n \cap Vs_m = \emptyset, \; m \neq n.$$

By induction we can get a sequence $\{f_n\} \subset C(G)$, a sequence $\{V_n\}$ of neighbourhoods of e and a sequence $\{t_n\} \subset G$ such that, for n = 1, 2, ..., we have (i)' $V_1 \subset V$, $V_{n+1} \subset V_n$.

(ii)' $f_n = 0$ off V_n, $f_n = 1$ on V_{n+1}, $0 \leq f_n \leq 1$.

(iii)' $t_n \in V_n \backslash V_{n+1}$.

For each n, define $h_n \in C(G)$ by $h_n(t) = f_n(ts_n^{-1})$ for all $t \in G$, and put $h = \sum h_n$. It follows from * and (ii)' that $h \in C(G)$. And, $h \notin WAP$ (which is the desired contradiction) since

$$h(t_m s_n) = f_n(t_m) = \begin{cases} 1 \text{ if } m > n \\ \\ 0 \text{ if } m < n \end{cases}$$

and hence $\lim_m \lim_n h(t_m s_n) = 0$ and $\lim_n \lim_m h(t_m s_n) = 1$, which is impossible for a function in WAP; see Theorem 8.2, (a) and (c).

14.13. Remark: For topological groups G, the equality $WAP = C(G)$ implies $AP = WAP$. One would not expect this to be the case for non-compact, semitopological semigroups; and, indeed, an example is provided in [Macri (1974)].

15. EXTENSION OF FUNCTIONS

In this section the problem under consideration is: if f is a function of a certain type on a subsemigroup S of a semitopological semigroup T, when does f extend to a function of the same type on T? We have a number of theorems and counterexamples, but no definitive answer to the question. We only state the first theorem; it is due to Katětov (1951).

15.1. Theorem: Let f be a bounded function uniformly continuous on a subspace Y of a uniform space X. Then f has a bounded uniformly continuous extension to X.

15.2. Corollary: Let G be a topological group with subgroup H. Then every function in *LUC*(H) extends to a function in *LUC*(G).

15.3. Remarks: Katětov's theorem can be applied to prove this corollary because, in the setting of topological groups, *LUC* is the set of bounded functions uniformly continuous with respect to the right uniformity (see Remark 5.2) which behaves "properly" with respect to the taking of subgroups; that is, if H is a subgroup of a topological group G, then the right uniformity of H, which H gets by virtue of being a topological group itself, is the same as the uniformity H gets as a subspace of G, G being furnished with its right uniformity.

This method of proof does not seem to apply to other extension theorems presented here. For example, *WAP*(T) is always a C*-algebra for any semitopological semigroup T and, if (w,WT) is the canonical *WAP*-compactification, then *WAP*(T) is precisely the subset of *C*(T) whose members are uniformly continuous with respect to the (unique) uniformity T gets by being the inverse image of the (compact) uniform space WT; if S is a subsemigroup of T, then the functions in *WAP*(S) all extend to functions in *WAP*(T) if and only if WS is isomorphically embedded in WT, i.e., if and only if the uniformity S gets by being the inverse image of WS is the same as the uniformity it gets from WT by taking inverse images using the composition

$$S \overset{i}{\to} T \overset{w}{\to} WT,$$

where i is the inclusion map. Examples in V.2 show this is not always the case.

15.4. Theorem [Milnes (1976)]: Let S be a dense subsemigroup of a topological semigroup G that is algebraically a group. Then every $f \in LMC(S)$ extends to a function in $C(G)$, i.e., $C(G)|_S \supset LMC(S)$.

Proof: If S is a dense subsemigroup of G and $r \in S$, then $T = rSr$ is a subsemigroup of S that is dense in G; for, the map $t \to rtr$ is a homeomorphism of G onto itself, since G is a group and a topological semigroup. We now produce a contradiction from the assumption that there exist $s \in G$ and nets $\{s'_\alpha\}$ and $\{t'_\beta\}$ in T such that

$$\lim_\alpha s'_\alpha = s = \lim_\beta t'_\beta$$

and

$$a = \lim_\alpha f(s'_\alpha) \neq \lim_\beta f(t'_\beta) = b$$

for an $f \in LMC(S)$. This will complete the proof.

Let $s_\alpha = r^{-1}s'_\alpha$ for each α and let $t_\beta = t'_\beta r^{-1}$ for each β. Then $\{s_\alpha\} \subset S$ and $\{t_\beta\} \subset S$. Without loss we may assume

$$\lim_\alpha e(s_\alpha) = \mu \in \beta S$$

(where e is the canonical continuous map of S into βS); thus $\mu(L_r f) = a$. Since T is dense in G, there is a net $\{r_\gamma\} \subset T$ such that

$$\lim_{\gamma} r_{\gamma} = s^{-1} \in G.$$

Now consider the values of f along the "triple net"

$$\{t_{\beta} r r_{\gamma} r s_{\alpha}\} \subset S.$$

Since $f \in LMC(S)$, the function $t \to \mu(L_t f)$ is continuous on S and, since

$$t_{\beta} r r_{\gamma} r \to r \in S$$

by continuity of multiplication, $\mu(L_{t_{\beta} r r_{\gamma} r} f)$ should approach $\mu(L_r f) = a$. But

$$\mu(L_{t_{\beta} r r_{\gamma} r} f) = \lim_{\alpha} f(t_{\beta} r r_{\gamma} r s_{\alpha})$$

is close to $f(t_{\beta} r) = f(t'_{\beta})$ for all large enough γ since $r_{\gamma} r s_{\alpha} \to e$. This implies that

$$\mu(L_{t_{\beta} r r_{\gamma} r} f) \to \lim_{\beta} f(t_{\beta} r) = b \neq a = \mu(L_r f)$$

as $t_{\beta} r r_{\gamma} r \to r$, the desired contradiction.

15.5. Remarks: In the familiar example of the rationals Q as a dense subgroup of the usual additive real numbers R, one can construct functions in $LMC(Q)$ whose continuous extensions to R are not in $LMC(R)$; see Example V.2.5. Examples V.2, 3 and 4, show that the conclusion of the theorem fails if the subsemigroup is not required to be dense or if the containing semigroup is not required to be a group.

15.6. Corollary: Let S be a subsemigroup of a compact topological group G. Then $LMC(G)|_S = LMC(S)$. Hence, $LMC(S) = AP(S) = SAP(S)$.

Proof: By [Hewitt and Ross (1963); (9.16) Theorem], the closure of S in G is a group. The results now follow from Theorems 15.4 and 15.1 and Remark 14.5 (iv).

15.7. <u>Corollary</u>: Let S be a dense subsemigroup of a topological semigroup G that is algebraically a group. Then $AP(G)|_S = AP(S)$ and $WAP(G)|_S = WAP(S)$.

Proof: Since $AP \subset WAP \subset LMC$, it follows from the theorem that we may regard $AP(S)$ and $WAP(S)$ as subsets of $C(G)$. For an $s \in G$, let $\phi(s)$ be the restriction of $e_G(s)$ to $AP(S)$ ($WAP(S)$). Then ϕ is a continuous homomorphism of G into AS (WS). It follows from Theorem 9.4 (8.4) that there is a continuous homomorphism of AG onto AS (WG onto WS), which implies $AP(G)|_S \supset AP(S)$ ($WAP(G)|_S \supset WAP(S)$). The reverse inclusion is easily established. (See Remark 2.8 (ii).)

Before stating some more corollaries, we list the following four theorems, the third of which is seen to be a consequence of the fact that, for a locally compact abelian group, AP is the closed linear span of the continuous characters [Hewitt and Ross (1970); 33.26 (g)], [Hewitt and Ross (1963); 24.12] and [Dixmier (1964); Corollaire 1.8.3].

15.8. <u>Theorem</u> [deLeeuw and Glicksberg (1965); Theorem 7.1]: Let H be a subgroup of a locally compact abelian group G. Then $WAP(G)|_H = WAP(H)$.

15.9. <u>Theorem</u> [Milnes (1975)]: Let H be an open subgroup of a locally compact group G. Then $WAP(G)|_H = WAP(H)$. In fact, the map ϕ defined for $f \in WAP(H)$ by $\phi f(s) = f(s)$,

$s \in H$, $\phi f(s) = 0$ otherwise, injects $\mathcal{WAP}(H)$ into $\mathcal{WAP}(G)$.

15.10. Theorem [Berglund (1970)]: Let H be a subgroup of a locally compact abelian group G. Then $\mathcal{AP}(G)|_H = \mathcal{AP}(H)$.

15.11. Theorem [Milnes (1975)]: Let H be a compact subgroup of a locally compact group G. Then $\mathcal{WAP}(G)|_H = \mathcal{WAP}(H)$.

15.12. Remark: Example V.2.6 shows that the analog of Theorem 15.11 for $\mathcal{AP}$ fails, even if H is also assumed to be open in G, while Example V.2.7 shows that the conclusion of Theorem 15.10 can fail if H is only required to be a subsemigroup of G.

15.13. Corollary: If S is a subsemigroup of a topological group G and S has compact closure in G, then $\mathcal{LUC}(G)|_S = \mathcal{LMC}(S)$. If, as well, G is locally compact, then $\mathcal{WAP}(G)|_S = \mathcal{LMC}(S)$.

15.14. Corollary: If S is a subsemigroup of a locally compact abelian group G and the closure of S in G is a group, then $\mathcal{AP}(G)|_S = \mathcal{AP}(S)$.

15.15. Corollary: If S is a subsemigroup of a locally compact group G, and the closure of S in G is an open subgroup of G, then $\mathcal{WAP}(G)|_S = \mathcal{WAP}(S)$.

15.16. Remark: The reader will observe from the examples in V.2 that most hypotheses in these results cannot be weakened. We note, however, that a fundamental result on the completion of topological groups [Weil (1937); p. 32] can be used to weaken some hypotheses. For example, the conclusion of

Theorem 15.11 still holds if the terms "compact group" and "locally compact group" are weakened to "totally bounded topological group" and "topological group with a totally bounded neighbourhood", respectively.

15.17. Theorem: Let S be a dense subsemigroup of a commutative topological semigroup T. Suppose that, for each $s \in T$, $sT \cap S \neq \emptyset$ (for example, if sT is open in T). Then $SAP(T)|_S = SAP(S)$.

Proof: Clearly $SAP(T)|_S \subset SAP(S)$. To show that the reverse inclusion holds, we show first that $SAP(S) \subset C(T)|_S$. Let $f \in SAP(S)$. It is enough to show that, if $\{s_\alpha\}$ and $\{t_\beta\}$ are nets in S converging to s such that

$$a = \lim_\alpha f(s_\alpha) \text{ and } b = \lim_\beta f(t_\beta)$$

exist, then a = b. Consider the canonical SAP-compactification (m,MS) of S. We may assume $m(s_\alpha) \to \mu$ and $m(t_\beta) \to \nu$ for some μ, $\nu \in MS$, so that $a = \mu(f)$ and $b = \nu(f)$. Choose $r \in T$ such that $sr \in S$ and let $\{r_\gamma\}$ be a net in S converging to r. Since MS is a group, there is a function $g \in SAP(S)$ such that $R_{sr}g = f$. Now, given $\varepsilon > 0$, choose indices β_0, γ_0 such that

$$|m(t_\beta rs)(g) - \nu m(rs)(g)| < \varepsilon, \; \beta \geq \beta_0,$$

and

$$|m(t_\beta r_\gamma)\mu(g) - m(sr)\mu(g)| < \varepsilon, \; \beta \geq \beta_0, \; \gamma \geq \gamma_0.$$

For fixed $\beta \geq \beta_0$, choose $\gamma \geq \gamma_0$ and α such that

$$|m(t_\beta r_\gamma s_\alpha)(g) - m(t_\beta)m(rs)(g)| < \varepsilon.$$

We may assume α also satisfies

$$|m(t_\beta r_\gamma s_\alpha)(g) - m(t_\beta r_\gamma)\mu(g)| < \varepsilon.$$

Then, since $\nu m(rs)(g) = b$ and $m(sr)\mu(g) = a$, it follows that $|a - b| < 4\varepsilon$. Thus $a = b$.

To show $SAP(T)|_S = SAP(S)$, let

$$F = \{f \in C(T) \mid f|_S \in SAP(S)\},$$

and define $V: F \to SAP(S)$ by $Vf = f|_S$. F is an m-introverted C*-subalgebra of $C(T)$ and V is an isometric isomorphism, which is surjective, since $C(T)|_S \supset SAP(S)$. Thus $V^*|_{MS}$ is an isomorphism of the topological group MS onto the canonical F-compactification $MM(F)$ of T. It follows that $MM(F)$ is a topological group and, hence, that $F \subset SAP(T)$. Thus $SAP(T)|_S = SAP(S)$.

15.18. We give some examples to which Theorem 15.17 applies.

(a) $T = [t,\infty)$ with ordinary addition $(t \geq 0)$, $S = T \cap Q$.

(b) $T = [0,t)$ with ordinary multiplication, where $0 < t \leq 1$ or $t = +\infty$, $S = T \cap Q$.

15.19. Lemma [Hildebrant and Lawson (1973)]: Let S be a right topological semigroup containing a dense left ideal I with $I \subset \Lambda(S)$. Suppose that T is a compact left reductive (i.e., $ta = tb$ for all $t \in T$ implies that $a = b$) right topological semigroup and that $\phi: I \to T$ is a continuous homomorphism with $\phi(I) \subset \Lambda(T) \subset \phi(I)^- = T$. Then there is a continuous homomorphism $\psi: S \to T$ such that the diagram

$$\begin{array}{ccc} I & \xrightarrow{i} & S \\ & \phi \searrow & \downarrow \psi \\ & & T \end{array}$$

commutes, where i is the inclusion map.

Proof: Fix $s \in S$. Suppose $\{x_\alpha\}$ and $\{y_\beta\}$ are nets on I with $\lim x_\alpha = s = \lim y_\beta$. Since T is compact, we may assume that $\phi(x_\alpha)$ and $\phi(y_\beta)$ converge to, say, u and v, respectively, in T. If $k \in I$, then, since $I \subset \Lambda(S)$, $kx_\alpha \to ks$. And since $\phi(I) \subset \Lambda(T)$, we get

$$\begin{aligned} \phi(ks) &= \lim \phi(kx_\alpha) = \lim \phi(k)\phi(xa) \\ &= \phi(k) \lim \phi(x_\alpha) = \phi(k)u. \end{aligned}$$

Likewise, $\phi(ks) = \phi(k)v$. Now, since T is right topological, and $\phi(I)$ is dense, we conclude that $tu = tv$ for all $t \in T$, which implies, since T is left-reductive, that $u = v$. It follows that ϕ extends to a continuous function on S.

15.20. Theorem [Hildebrant and Lawson (1973)]: Let S be a semitopological semigroup containing a dense left ideal I. Let H denote any one of P, K, U, W, A, M, Q, or V. If the compact semigroup HI is left reductive, then HI is topologically isomorphic to HS.

Proof: This is a direct result of Lemma 15.19 and the appropriate universal mapping property.

15.21. Corollary: Let S be a semitopological semigroup containing a dense left ideal I. Suppose that I contains a left identity element. Then

$$LMC(I) = LMC(S)|_I,\ K(I) = K(S)|_I,$$
$$LUC(I) = LUC(S)|_I,\ WAP(I) = WAP(S)|_I,$$
$$AP(I) = AP(S)|_I,\ SAP(I) = SAP(S)|_I,$$
$$KWP(I) = KWP(S)|_I, \text{ and } LWP(I) = LWP(S)|_I.$$

15.22. Remark: Analogous extension theorems may be obtained for $WLUC(I)$, $CK(I)$, and $CKWP(I)$.

16. DIRECT SUMS OF SUBSPACES OF $C(S)$

Throughout this section S denotes a semitopological semigroup, F is a translation invariant, left m-introverted C*-subalgebra of $C(S)$ containing the identity, and (ψ, X) is the canonical F-compactification of S (2.2). We shall analyse the ideal structure of F and determine conditions under which F is the direct sum of two subspaces, one of which is a C*-subalgebra of "reversible functions", the other an ideal of "flight functions". The material in this section is a generalization and elaboration of results of deLeeuw and Glicksberg on the splitting of $\mathcal{WAP}(S)$. It depends in an essential way on the structure theory developed in Chapter II.

For this section only we shall need the following definition of left m-introverted, which is more general than that given in I.4.11.

16.1. Definition: A subset H of F is left m-introverted if $T_x H \subset H$ for all $x \in X$, where T_x: $F \to F$ is the map defined by

$$(T_x f)(s) = x(L_s f), \ f \in F, \ s \in S.$$

We remind the reader that if $H \subset F$ and $J \subset F^*$ then $H^\perp$ and $J^\perp$ denote, respectively, the annihilators of H and J relative to the duality (F,F^*). That is,

$$H^\perp = \{x \in F^* \mid x(H) = \{0\}\},$$
$$J^\perp = \{f \in F \mid J(f) = \{0\}\}.$$

16.2. Definition: If H is a subset of F we denote by I_H the $\sigma(F^*,F)$ closed set $X \cap H^\perp$.

The following lemma particularizes to our setting the well-known result that $H \to I_H$ defines a one-to-one correspondence between norm closed, conjugate closed ideals of F and closed subsets of X.

16.3. Lemma: Let H and H_1 be conjugate closed, norm closed, proper ideals of F.

(i) $I_H \neq \emptyset$ and $I_H^{\perp} = H$.

(ii) I_H is a left ideal of X if and only if H is left translation invariant.

(iii) I_H is a right ideal of X if and only if H is left m-introverted.

(iv) $I_H \subset I_{H_1}$ if and only if $H_1 \subset H$.

(v) If J is a closed left (respectively, right) ideal of X, then $J^{\perp}$ is a left translation invariant (respectively, left m-introverted), conjugate closed, norm closed ideal in F, and $I_{J^{\perp}} = J$.

Proof: (i) Let Y denote the spectrum of the quotient algebra F/H and $Q: F \to F/H$ the quotient mapping. Then $Q^*Y = I_H$, hence $I_H^{\perp} = \ker Q = H$.

(ii) This follows from (i) and the observation that I_H is a left ideal of X if and only if, for each $s \in S$ and $x \in I_H$,

$$L_s^* x = \psi(s)x \in I_H.$$

(iii) This follows from (i) and definition of multiplication in X (I.4.13).

(iv) This follows from properties of annihilators.

(v) This follows directly from (i), (ii), and (iii).

16.4. Remarks: (i) Parts (ii) and (iv) of the lemma combine to show that I_H is a minimal left ideal of X if and only if H is a maximal, left translation invariant, conjugate closed, proper ideal of F. Similar remarks apply with regard to I_H

being a minimal closed right ideal or a minimal closed two-sided ideal of X. In particular, since the latter is unique, F has a largest (in the sense of containment) left translation invariant, left m-introverted, conjugate closed, proper ideal.

(ii) Suppose F is left amenable and $\mu \in LIM(F)$. Then $H = \{f \in F \mid \mu(|f|) = 0\}$ is a left translation invariant, norm closed, conjugate closed, proper ideal of F and, from 16.3 (i), $I_H = \text{supp}\ \mu$; see II.5. Thus, if $F \subset WAP(S)$, then H is a maximal, left translation invariant, conjugate closed, proper ideal of F if and only if μ is an extreme left invariant mean (II.5.7).

16.5. Definition: Let p denote the topology on $C(S)$ of pointwise convergence and, for $f \in F$, let $p - \text{cl}\ R_S f$ be the p-closure of $R_S f$. We define

$$F_0 = \{f \in F \mid 0 \in p - \text{cl}\ R_S f\},$$

and also we define F_d to be the largest, left translation invariant, left m-introverted, conjugate closed, proper ideal of F (discussed in 16.4 (i)).

16.6. Remarks: (i) If $F = WAP(S)$ then F_0 coincides with the set of "flight vectors" in the sense of [Jacobs (1956)], and F_d is the "dissipative" subspace of [Berglund and Hofmann (1967)]. Also, F_d corresponds to the set X_z of Theorem II.1.33.

(ii) It is readily verified that F_0 has all the properties of a left translation invariant, norm closed, conjugate closed, proper ideal of F except, possibly, closure under addition. Part (iii) of Lemma 16.7 below tells us precisely when F_0 possesses this latter property.

16.7. Lemma:

(i) $F_0 = \cup \{\ker T_x \mid x \in X\} = \cup \{\ker T_e \mid e \in E(K(X))\}$.

(ii) F_0 contains every left translation invariant, conjugate closed, proper ideal of F.

(iii) F_0 is closed under addition if and only if X has a unique minimal left ideal. In this case $F_0 = F_d = \ker T_e$ for any $e \in E(K(X))$, and $I_{F_0} = K(X)$.

Proof: (i) The first equality follows immediately from the fact that $p - \mathrm{cl}\, R_S f = T_x f$ (I.4.19). For the second, simply note that, if $T_x f = 0$, then the set $\{y \in X \mid T_y f = 0\}$ is a left ideal and therefore meets $E(K(X))$ (II.2.2).

(ii) Let H be any left translation invariant, norm closed, conjugate closed, proper ideal of F, and choose any $x \in I_H$. Then, if $f \in H$ and $s \in S$, $(T_x f)(s) = x(L_s f) = 0$, hence $f \in \ker T_x \subset F_0$.

(iii) If F_0 is closed under addition, then by (ii) and 16.6 (ii) F_0 is the largest left translation invariant, conjugate closed, proper ideal of F. Hence I_{F_0} is the unique minimal left ideal of X (16.4 (i)). Conversely, if X has a unique minimal left ideal, then for any $e_1, e_2 \in E(K(X))$ $\ker T_{e_1} = \ker T_{e_2}$ (because $e_i e_j = e_i$ (II.2.2)), hence F_0 is closed under addition.

16.8. Definition: $F_r = \{f \in F \mid g \in p - \text{cl } R_S f$ implies $f \in p - \text{cl } R_S g\}$. If $F = WAP(S)$ then F_r is the set of "reversible vectors" in the sense of [Jacobs (1956)]. (See also II.1.30 (i).)

16.9. Lemma: (i) $F_r = \cup \{\ker (T_x - I) \mid x \in K(X)\}$

$$= \cup \{\ker (T_e - I) \mid e \in E(K(X))\},$$

where I is the identity operator.

(ii) X has a unique minimal right ideal if and only if F_r is a left translation invariant, left m-introverted C*-subalgebra of $C(S)$ and the F_r- compactification of S is a compact right topological group. In this case, $(\rho_e \psi, Xe)$ is an F_r-compactification of S for any $e \in E(K(X))$ (where ρ_e: $X \to Xe$ is defined by $\rho_e x = xe$).

Proof: (i) Let $f \in F_r$ and $e \in E(K(X))$. Then

$$T_e f \in T_X f = p - \text{cl } R_S f$$

implies that $f \in T_{Xe} f$; hence $Y = \{x \in Xe \mid T_x f = f\}$ is a non-empty compact subsemigroup of X, so $E(Y) \neq \emptyset$. Therefore $f \in \ker(T_{e_1} - I)$ for some $e_1 \in E(K(X))$. The same argument also shows that $\ker(T_x - I)$ is contained in

$$\cup \{\ker(T_e - I) \mid e \in E(K(X))\}$$

for each $x \in K(X)$. Now let $f \in \ker(T_e - I)$, where $e \in E(K(X))$. If $g \in p - \text{cl } R_S f$, then $g = T_x f$ for some $x \in X$. By minimality $Xxe = Xe$, hence there exists $y \in X$ such that $yxe = e$. Therefore

$$T_y g = T_{yxe} f = T_e f = f,$$

so $f \in F_r$.

(ii) Suppose X has a unique minimal right ideal, and let $e, e_1 \in E(K(X))$. Then, from $ee_1 = e_1$ and $e_1e = e$ (II.2.2), it follows that $\ker(T_e - I) = \ker(T_{e_1} - I)$; hence $F_r = \ker(T_e - I)$. Since $\ker(T_e - I) = (\rho_e\psi)^*C(Xe)$, the "only if" part of (ii) is established.

Conversely, let F_r have the stated properties and let (γ, Y) be an F_r-compactification of S, where Y is a compact right topological group. Then, for any $e \in E(K(X))$, $\ker(T_e - I) = F_r$. For, if $\theta: X \to Y$ is the unique continuous homomorphism such that $\theta\psi = \gamma$, then $\theta(e)$ is the identity of Y and consequently, for any $f \in F_r$,

$$T_e f = \psi^* R_e \psi^{*-1} f = \psi^* R_e \theta^* (\gamma^*)^{-1} f = \psi^* \theta^* (\gamma^*)^{-1} f = f.$$

Thus $(\rho_e\psi)^*C(Xe) = \ker(T_e - I) = F_r = \gamma^*C(Y)$, hence θ, which is a homomorphism of the minimal left ideal Xe onto Y is also 1-1 on Xe. This implies that K(X) is a minimal right ideal as desired.

16.10. <u>Corollary</u> [deLeeuw and Glicksberg (1961)]: If $F \subset WAP(S)$, then $F_r = F \cap SAP(S)$ if and only if F is left amenable.

Proof: 16.9 (ii), II.5.6, and 10.6.

Theorems 16.12 and 16.13 below are the main results of this section. The following lemma, which is of some interest in itself, provides the essential step in the proofs of those theorems.

16.11. <u>Lemma</u>: Let H be a norm closed, conjugate closed, proper ideal of F. Then $F = G \oplus H$ for some C*-subalgebra G containing the identity if and only if I_H is a (topological) retract of X.

Proof: If $F = G \oplus H$ and P: $F \to G$ is the canonical projection then P*R: X $\to$ X is a retraction onto I_H, where R: X $\to$ MM(G) denotes the restriction map.

Conversely, if ϕ: X $\to I_H$ is a retraction onto I_H, then $F = G \oplus H$, where $G = \psi^*\phi^*C(I_H)$.

16.12. <u>Theorem</u>: If X has a unique minimal left ideal, then $F = G \oplus F_0$, where G is the unique left m-introverted, translation invariant C*-subalgebra of F contained in F_r and complemented by a left translation invariant, norm closed, conjugate closed ideal in F. If X also has a unique minimal right ideal then $F = F_r \oplus F_0$.

Proof: Recall that $F_0 = F_d$ in this setting (16.7 (iii)). (Also G corresponds to the set X_g of Theorem II.1.28.) Let $e \in E(K(X))$. Then ρ_e: X $\to$ Xe = K(X) is a retraction; hence, by 16.11 and 16.7 (iii), $F = G \oplus F_0$, where

$$G = \psi^*\rho_e^*C(Xe) = \ker(T_e - I) \subset F_r,$$

the latter inclusion by 16.9 (i). Now suppose $F = G_1 \oplus H$, where G_1 is a left m-introverted C*-subalgebra of F contained in F_r, and H is a left translation invariant, norm closed, conjugate closed ideal in F. Let $g \in G$, $g = g_1 + h$, where $g_1 \in G_1$ and $h \in H$. Then $g - g_1 \in H \subset F_0$, so $T_e(g - g_1) = 0$, and $g = T_e g = T_e g_1 \in G_1$. A similar argument shows that $G_1 \subset G$.

If X also has a unique minimal right ideal then $G = F_r$ by 16.9 (ii).

16.13. Theorem: If X has a unique minimal right ideal and $F \subset WAP(S)$, then $F = G \oplus H$, where $G = F \cap SAP(S)$ and $H = \{f \in F \mid \mu(|f|) = 0\}$ for some extreme left invariant mean μ on F.

Proof: Let $e \in E(K(X))$. Since $\rho_e: X \to Xe$ is a retraction it follows from 16.3 (v) and 16.11 that $F = G \oplus (Xe)^{\perp}$, where $G = \psi^*\rho_e^*C(Xe) = T_eF = F \cap SAP(S)$, the last equality by Theorem 10.6. Let ν be normalized Haar measure on the compact group Xe, and define $\mu \in LIM(F)$ by

$$\mu(f) = \int_{Xe} \hat{f}(x)\nu(dx), \quad f \in F, \ \psi^*(\hat{f}) = f.$$

If $H = \{f \in F \mid \mu(|f|) = 0\}$ then by 16.4 (ii), $I_H = \operatorname{supp} \mu = Xe$, so $H = I_H^{\perp} = (Xe)^{\perp}$.

16.14. Corollary [deLeeuw and Glicksberg (1961)]: If $WAP(S)$ is amenable, then $WAP(S) = SAP(S) \oplus WAP(S)_0$.

CHAPTER IV

FIXED POINTS AND LEFT INVARIANT MEANS ON SUBSPACES OF $C(S)$

In this chapter we show that the existence of fixed points of various types of flows (X,S) is equivalent to the existence of left invariant means on certain subspaces of $C(S)$ (where S is a semitopological semigroup, as usual). The first such general result appeared in a paper of M. M. Day (1961) and yielded as a simple corollary the Markov-Kakutani Fixed Point Theorem. Since then many fixed point theorems of this type have appeared in the literature, and in the material that follows we attempt to give a unified presentation of some of these results.

The first section relates the existence of fixed points of affine flows (X,S) to the existence of left invariant means (LIM's) on subspaces of $C(S)$. In the second section we demonstrate the connection between the existence of fixed points of general flows (X,S) and the existence of multiplicative left invariant means (MLIM's) on subalgebras of $C(S)$.

1. FIXED POINTS OF AFFINE FLOWS AND LEFT INVARIANT MEANS

The following lemma, which is used to prove Theorem 1.3 below, is of some independent interest.

1.1. Lemma: Let (X,S,π) be an affine flow (see I.2.1) and let $\phi: S \to X$ be a continuous map such that

(1) $$\phi(st) = \pi(s,\phi(t)), \quad s, t \in S.$$

Let F be a left translation invariant, conjugate closed, norm closed, linear subspace of $C(S)$ containing the constant functions, and let $e: S \to M(F)$ denote the evaluation map. Then $\phi^*A(X) \subset F$ if and only if there exists a continuous affine mapping $\psi: M(F) \to X$ such that $\psi \circ e = \phi$. In this case ψ satisfies

$$(2) \qquad \psi(L_s^*\mu) = \pi(s,\psi(\mu)), \quad s \in S, \ \mu \in M(F).$$

Proof: Suppose $\phi^*A(X) \subset F$. Define $\psi: \mathrm{coe}(S) \to X$ by

$$\psi\left(\sum_{s\in S} a(s)e(s)\right) = \sum_{s\in S} a(s)\phi(s).$$

Using the fact that $e^*A(M(F)) = F$ (I.3.7) and the hypothesis $\phi^*A(X) \subset F$, one shows (as in the proof of III.1.4) that ψ is well defined and has a continuous extension to $M(F)$. Since $e(st) = L_s^*e(t)$, we have from (1) that

$$\psi(L_s^*e(t)) = \pi(s,\psi(e(t))).$$

Equation (2) then follows from the continuity of the maps

$$\mu \to \psi(L_s^*\mu) \text{ and } \mu \to \pi(s,\psi(\mu)).$$

On the other hand, if such a map ψ exists, then

$$\phi^*A(X) = e^* \circ \psi^*A(X) \subset e^*A(M(F)) = F.$$

1.2. Definition: Let (X,S) be a flow. For each $x \in X$ define the map $T_x: C(S) \to B(S)$ by

$$(T_x h)(s) = h(sx), \quad s \in S, \ h \in C(X).$$

1.3. Theorem [Argabright (1968), Day (1961)]: Let F be a translation invariant, conjugate closed, norm closed, linear subspace of $C(S)$ containing the constant functions. Then conditions (a), (b) and (c) below are equivalent and imply (d). If F is left introverted then (a) - (d) are all equivalent.

(a) F is left amenable.

(b) Every affine flow (X,S,π) such that $\{x \in X \mid T_x A(X) \subset F\} \neq \emptyset$ has a fixed point.

(c) Every affine flow (X,S,π) such that $\{x \in X \mid T_x A(X) \subset F\}$ is dense in X has a fixed point.

(d) Every affine flow (X,S,π) such that $\{x \in X \mid T_x A(X) \subset F\} = X$ has a fixed point.

Proof: (a) implies (b). Let (X,S,π) be an affine flow such that $\{x \in X \mid T_x A(X) \subset F\} \neq \emptyset$, and choose one such x. Define $\phi: S \to X$ by $\phi(s) = sx$. Then ϕ is continuous and satisfies (1), hence there exists $\psi: M(F) \to X$ such that (2) holds. Therefore, if $\mu \in LIM(F)$, then $\psi(\mu)$ is a fixed point of (X,S,π).

It is obvious that (b) implies (c) and (c) implies (d).

(c) implies (a). Let $X = M(F)$ and define

$$\pi(s,\mu) = L_s^*\mu, \quad s \in S, \ \mu \in M(F).$$

Then (X,S,π) is an affine flow (I.4.4) and, since

$$T_{e(s)}h = R_s e^*h, \quad h \in A(X), \ s \in S,$$

it follows that $T_x A(X) \subset F$ for all $x \in coe(S)$. Hence, if (c) holds, then (X,S,π) has a fixed point, which is a LIM for F.

Now suppose F is left introverted, and let (X,S,π) be as in the proof that (c) implies (a). Then X is a compact affine right topological semigroup (I.4.14) and

$$T_x h = e^*R_x h, \quad h \in A(X), \ x \in X.$$

Therefore $T_x A(X) \subset F$ for all $x \in X$; so, if (d) holds (X,S,π) has a fixed point, and consequently F is left amenable.

1.4. <u>Remark</u>: If $X = M(F)$ and $\pi(s,\mu) = L_s^*\mu$, then the condition in (d) that $T_x A(X) \subset F$ for all $x \in X$ is equivalent to the assertion that F is left introverted.

1.5. <u>Definition</u>: Let (X,S) be an affine flow with separately continuous action, and let τ be a locally convex topology on $C(S)$. Then (X,S) is called a <u>τ-affine flow</u> if

$$x \to T_x h: X \to C(S)$$

is τ-continuous for each $h \in A(X)$.

1.6. <u>Theorem</u> [Junghenn (1975)]: Let τ be a locally convex topology on $C(S)$ such that $p \leq \tau \leq u$, where p and u denote, respectively, the pointwise and norm topologies on $C(S)$. Suppose the maps $f \to f^*$ and L_t, $t \in S$, from $C(S)$ into $C(S)$ are τ-continuous. Then

$$F = \{f \in C(S) \mid \mathrm{co}R_S f \text{ is relatively } \tau\text{-compact}\}$$

is left amenable if and only if every τ-affine flow (X,S) has a fixed point.

Proof: Let F be left amenable and let (X,S) be a τ-affine flow. If y is any member of X and $h \in A(X)$, then

$$\mathrm{co}R_S T_y h = T_Y h, \tag{3}$$

where $Y = \mathrm{co}Sy$. Since $x \to T_x h$ is τ-continuous, $T_Y h$ is relatively τ-compact; hence, by (3), $T_y h \in F$. Therefore (X,S) has a fixed point by 1.3.

Conversely, let $X = M(F)$ with its relativized $\sigma(C(S)^*, C(S))$ topology. Then (X,S) is an affine flow under $(s,\mu) \to L_s^*\mu$. Furthermore, (X,S) is τ-affine, since, if $h \in A(X)$, then the p-continuity of $\mu \to T_\mu h: X \to C(S)$ and the relative τ-compactness of

$$T_{\mathrm{coe}(S)}h = \mathrm{co}R_S e^*h$$

imply that $\mu \to T_\mu h$ is τ-continuous. Hence, by assumption, (X,S) has a fixed point, which is obviously a LIM for F.

1.7. Remark: A slightly more general version of Theorem 1.6 may be obtained by replacing the single topology τ by a family of such topologies. We leave the precise formulation and proof to the reader. (See III.1.7.)

1.8. Definition: Let (X,S,π) be a flow with enveloping semigroup E. We shall say that (X,S,π) is

(a) separately continuous if $\pi: S \times X \to X$ is separately continuous.

(b) jointly continuous if π is continuous.

(c) equicontinuous if $\pi^S = \{\pi^s \mid s \in S\}$ is an equicontinuous family of mappings (with respect to the unique uniformity on X).

(d) quasi-equicontinuous if each member of E is continuous.

(e) equicontinuous on compacta if π^K is an equicontinuous family for each compact set $K \subset S$.

1.9. Theorem: (a) [Mitchell (1970)] $WLUC(S)$ is left amenable if and only if every separately continuous affine flow (X,S) has a fixed point.

(b) [Mitchell (1970)] $LUC(S)$ is left amenable if and only if every jointly continuous affine flow (X,S) has a fixed point.

(c) [Junghenn (1975)] $CK(S)$ is left amenable if and only if every separately continuous affine flow (X,S) which is equicontinuous on compacta has a fixed point.

(d) [Junghenn (1975), Lau (1976)] $\mathcal{WAP}(S)$ is left amenable if and only if every separately continuous quasi-equicontinuous affine flow (X,S) has a fixed point.

(e) [Lau (1973)] $\mathcal{AP}(S)$ is left amenable if and only if every separately continuous equicontinuous affine flow (X,S) has a fixed point.

(f) $\mathcal{LWP}(S)$ is left amenable if and only if every jointly continuous quasi-equicontinuous affine flow (X,S) has a fixed point.

(g) $\mathcal{CKWP}(S)$ is left amenable if and only if every separately continuous quasi-equicontinuous affine flow (X,S) which is equicontinuous on compacta has a fixed point.

Proof: (a) This follows immediately from 1.6 and III.3.2.

(b) If (X,S) is a jointly continuous affine flow, then, using I.1.8 (a), one easily shows that $T_xA(X) \subset \mathcal{LUC}(S)$ for all $x \in X$. Hence, if $\mathcal{LUC}(S)$ is left amenable, then every jointly continuous affine flow (X,S) has a fixed point by 1.3.

The converse follows from the fact that $(M(\mathcal{LUC}),S)$ is a jointly continuous affine flow under the action $(s,\mu) \to L_s^*\mu$.

(c) This follows immediately from 1.6 and the fact that a separately continuous affine flow (X,S) is a compact-open-affine flow if and only if it is equicontinuous on compacta.

(d) This result will follow from 1.6 if we show that a separately continuous affine flow (X,S,π) is quasi-equicontinuous if and only if $x \to T_xh\colon X \to \mathcal{C}(S)$ is $\sigma(\mathcal{C}(S),\mathcal{C}(S)^*)$ continuous for each $h \in A(X)$.

Suppose first that (X,S,π) is quasi-equicontinuous. Then its enveloping semigroup E is semitopological (I.2.2); so, if (ψ,Y) is a $\mathcal{WAP}$-compactification of S, there exists a continuous homomorphism $\phi: Y \to E$ such that $\phi(\psi(s)) = \pi^s$, $s \in S$ (III.8.4). Let $h \in A(X)$ and set

$$(T_x h)^\wedge = (\psi^*)^{-1}(T_x h),\ x \in X.$$

Then $(T_x h)^\wedge(\psi(s)) = h(sx) = h(\pi^s x) = h(\phi(\psi(s))(x))$; hence, by continuity,

$$(T_x h)^\wedge(y) = h(\phi(y)(x)),\ y \in Y.$$

The map $x \to (T_x h)^\wedge: X \to C(Y)$ is therefore continuous if $C(Y)$ has the topology of pointwise convergence. By a theorem of Grothendieck (1952), $x \to (T_x h)^\wedge$ is actually $\sigma(C(Y),C(Y)^*)$ continuous, and therefore $x \to T_x h: X \to C(S)$ is $\sigma(C(S),C(S)^*)$ continuous.

Conversely, assume that $x \to T_x h: X \to C(S)$ is $\sigma(C(S),C(S)^*)$ continuous for each $h \in A(X)$, and let $f \in E$. If $\{\pi^{s_\alpha}\}$ is a net in π^S converging pointwise to f, then, for any $x \in X$ and $h \in A(X)$,

(4) $$(h \circ f)(x) = \lim_\alpha (T_x h)(s_\alpha).$$

Let $e: S \to C(S)^*$ denote the evaluation map and let $\{s_\beta\}$ be a subnet of $\{s_\alpha\}$ such that $\{e(s_\beta)\}$ $\sigma(C(S)^*,C(S))$ converges to $\mu \in C(S)^*$. Then from (4),

$$(h \circ f)(x) = \mu(T_x h),$$

hence $h \circ f$ is continuous. Since $h \in A(X)$ was arbitrary, f is continuous. Therefore (X,S) is quasi-equicontinuous.

(e) By I.1.8 (a), a separately continuous affine flow (X,S) is equicontinuous if and only if it is a norm-affine flow. The result now follows from 1.6

(f) Combine the proofs of (b) and (d), and use 1.6.

(g) Combine the proofs of (c) and (d), and use 1.6.

1.10. <u>Remarks</u>: (a) If G is the (discrete) free group on two generators, then $LUC(G) = B(G)$ is not left amenable [von Neumann (1929)]. (A proof of this result is given in [Greenleaf (1969)], as is a thorough discussion of amenable groups in general; worthy of note is the still not completely resolved conjecture of von Neumann: every non-amenable locally compact group contains a copy of the free group on two generators as a closed subgroup.) On the other hand, $B(S)$ is amenable for any abelian semigroup S. Also, $WAP(G)$ is amenable for any group G (1.14 ahead), and $SAP(S)$ is amenable for any semigroup S, the (unique) invariant mean on SAP being given by integration with respect to Haar measure on a SAP-compactification of S (which is a compact topological group).

(b) If G is a locally compact group, then the following assertions about G are equivalent.

(i) $LUC(G)$ is left amenable.

(ii) $C(G)$ is left amenable.

(iii) $L^\infty(G)$ is left amenable.

See [Greenleaf (1969); Theorem 2.2.1].

(c) $AP(S)$ is left amenable for all <u>left reversible</u> semitopological semigroups S (i.e., for all semitopological semigroups S with the property that any pair of non-empty closed right ideals in S have a non-empty intersection) [Lau (1973); Corollary 3.3].

(d) If S is a left zero semigroup with two elements ($st = s$, $s, t \in S$), then $AP(S) = C(S)$ is not left amenable.

For our next result we shall require the following lemma,

due to Granirer and Lau.

1.11. Lemma [Granirer and Lau (1971)]: Let F be a translation invariant, left introverted, conjugate closed, norm closed subspace of $C(S)$ containing the constant functions. Then F is left amenable if and only if, for each $f \in F$ and each $t \in S$, there exists $\mu \in M(F)$ such that $T_\mu(f - L_t f) = 0$. (See I.4.9 for the definition of T_μ.)

Proof: The proof in the forward direction is obvious. Conversely, for each $f \in F$, $t \in S$, let

$$K(f,t) = \{\nu \in M(F) \mid T_\nu(f - L_t f) = 0\}.$$

The sets $K(f,t)$ are $\sigma(F^*,F)$ compact and, by hypothesis, nonempty. We shall show that the family $\{K(f,t) \mid f \in F, t \in S\}$ has the finite intersection property. From this we can conclude that there exists $\nu \in M(F)$ such that $T_\nu f = T_\nu L_t f$ for all $f \in F$ and all $t \in S$. Then $\nu^2 = \nu \circ T_\nu \in LIM(F)$.

Let $f_1, f_2, \ldots, f_n \in F$ and $t_1, t_2, \ldots, t_n \in S$. By induction, we may assume that there exists

$$\mu \in \bigcap_{i=1}^{n-1} K(f_i,t_i).$$

Choose any $\nu \in K(T_\mu f_n, t_n)$. Then, by I.4.10 and I.4.14 (a)

$$T_{\nu\mu}(f_n - L_{t_n} f_n) = T_\nu(T_\mu f_n - L_{t_n} T_\mu f_n) = 0,$$

and, if $1 \leq i \leq n - 1$,

$$T_{\nu\mu}(f_i - L_{t_i} f_i) = T_\nu T_\mu(f_i - L_{t_i} f_i) = T_\nu 0 = 0.$$

Therefore $\nu\mu \in \bigcap_{i=1}^{n} K(f_i,t_i)$.

1.12. Theorem [Granirer and Lau (1971)]: Let p denote the topology on $C(S)$ of pointwise convergence, and let F be a translation invariant, left introverted, conjugate closed, norm closed subspace of $C(S)$ containing the constant functions. Then the following are equivalent:

(a) F is left amenable.

(b) The p-closure of $\mathrm{co}R_S f$ contains a constant function for each $f \in F$.

(c) The p-closure of $\mathrm{co}R_S(f - L_t f)$ contains 0 for each $t \in S$, $f \in F$.

Moreover, if F is left amenable, then, for each $f \in F$,

$$K(f) = \{\mu(f) \mid \mu \in \mathrm{LIM}(F)\},$$

where $K(f) = \{a \mid a1$ is a constant function in the p-closure of $\mathrm{co}R_S f\}$.

Proof: (a) implies (b). If (a) holds and $\mu \in \mathrm{LIM}(F)$, then the constant function $T_\mu f\colon t \to \mu(f)$ is in the p-closure of $\mathrm{co}R_S f$, $f \in F$. Thus (b) holds and

$$K(f) \supset \{\mu(f) \mid \mu \in \mathrm{LIM}(F)\}.$$

To show the reverse inclusion, let $c \in K(f)$. Then, by I.4.19, there exists $\nu \in M(F)$ such that $T_\nu f = c$. Choose any $\mu \in \mathrm{LIM}(F)$. Then $\mu\nu \in \mathrm{LIM}(F)$ (I.4.15 (b)) and $(\mu\nu)(f) = c$.

(b) implies (c). Let $f \in F$, $t \in S$. If c is in the p-closure of $\mathrm{co}R_S f$, let $\nu \in M(F)$ be such that $T_\nu f = c$. Then

$$0 = T_\nu f - L_t T_\nu f = T_\nu(f - L_t f)$$

is in the p-closure of $\mathrm{co}R_S(f - L_t f)$.

(c) implies (a). Suppose (c) holds, and let $f \in F$, $t \in S$. By I.4.19 there exists $\mu \in M(F)$ such that $T_\mu(f - L_t f) = 0$. Therefore, by Lemma 1.11, F is left amenable.

1.13. Remarks: (a) If $\mathcal{F} \subset \{f \in \mathcal{C}(S) \mid \mathrm{co}R_S f$ is relatively τ-compact$\}$, where τ is as in 1.6, then, for each $f \in \mathcal{F}$, the p- and τ-closures of relatively τ-compact subsets of $\mathcal{C}(S)$ coincide. Hence, in this case, the topology p in the statement of 1.12 may be replaced by τ. For example, if $\mathcal{F} = \mathcal{LUC}(S)$, then we may take τ to be the compact-open topology (since $\mathcal{LUC}(S) \subset \mathcal{CK}(S)$).

(b) If S is discrete and $\mathcal{F} = \mathcal{B}(S)$, then 1.12 reduces to Theorem 3 of [Mitchell (1965)].

1.14. Corollary [Ryll-Nardzewski]: If G is a group and a semitopological semigroup, then $\mathcal{WAP}(G)$ has a unique (left and right) invariant mean μ.

Proof: By the Ryll-Nardzewski Fixed Point Theorem (see [Berglund and Hofmann (1967)] for a self-contained geometric proof due to I. Namioka and E. Asplund), the weak closure of $\mathrm{co}R_S f$ (which is the same as the p-closure of $\mathrm{co}R_S f$) contains a constant function for each $f \in \mathcal{WAP}(G)$. Therefore, by 1.12, $\mathcal{WAP}(G)$ has a LIM μ. Similarly, $\mathcal{WAP}(G)$ has a right invariant mean ν. By I.4.15 (b) (and its "right" analog), $\mu = \nu\mu = \nu$. This argument also shows that μ is unique.

1.15. Remarks: (a) The uniqueness of μ implies that μ is inversion invariant. (If $f \in \mathcal{WAP}(G)$, define $\tilde{f} \in \mathcal{WAP}(G)$ by $\tilde{f}(s) = f(s^{-1})$; then $\nu(f) = \mu(\tilde{f})$ defines a (left and right) invariant mean ν, hence $\nu = \mu$.)

(b) The mean μ of 1.14 has the following property: if $f \in \mathcal{AP}(G)$, $f \geq 0$ and $f \neq 0$, then $\mu(f) > 0$. Indeed, $\mu|_{\mathcal{AP}(G)}$ is given by integration with respect to Haar measure on the compact group AG. See [deLeeuw and Glicksberg (1961); Theorem 5.8], Remarks 1.10, Corollary II.5.8, Theorem III.10.6 and Remarks III.8.9 in this regard.

2. FIXED POINTS OF FLOWS AND MULTIPLICATIVE LEFT INVARIANT MEANS

The results of this section are analogous to the results of section 1, hence we omit proofs.

2.1. Lemma: Let (X,S,π) be a flow and let $\phi: S \to X$ be a continuous map such that

$$\phi(st) = \pi(s,\phi(t)), \quad s, t \in S.$$

Let F be a left translation invariant C*-subalgebra of $C(S)$ containing the constant functions, and let $e: S \to MM(F)$ be the evaluation map. Then $\phi^*C(X) \subset F$ if and only if there exists a continuous mapping $\psi: MM(F) \to X$ such that $\psi \circ e = \phi$. In this case, ψ satisfies

$$\psi(L_s^*\mu) = \pi(s,\psi(\mu)), \quad s \in S, \ \mu \in MM(F).$$

2.2. Theorem [Mitchell (1968)]: Let F be a left translation invariant C*-subalgebra of $C(S)$ containing the constant functions. Then conditions (a), (b) and (c) below are equivalent and imply (d). If F is left m-introverted then (a) - (d) are all equivalent.

(a) F is extremely left amenable (i.e., has a multiplicative left invariant mean).

(b) Every flow (X,S) such that $\{x \in X \mid T_xC(X) \subset F\} \neq \emptyset$ has a fixed point.

(c) Every flow (X,S) such that $\{x \in X \mid T_x C(X) \subset F\}$ is dense in X has a fixed point.

(d) Every flow (X,S) such that $\{x \in X \mid T_x C(X) \subset F\}$ $= X$ has a fixed point.

2.3. Remark: If $X = MM(F)$ and $\pi(s,\mu) = L_s^*\mu$, then the condition in (d) that $T_x C(X) \subset F$ for all $x \in X$ is simply the assertion that F is left m-introverted.

2.4. Definition: Let (X,S) be a separately continuous flow and let τ be a locally convex topology on $C(S)$. Then (X,S) is called a τ-flow if $x \to T_x h: X \to C(S)$ is τ-continuous for each $h \in C(X)$.

2.5. Theorem [Junghenn (1975)]: Let τ be a locally convex topology on $C(S)$ such that $p \leq \tau \leq u$ and such that the maps $f \to f^*$ and L_t, $t \in S$, from $C(S)$ into $C(S)$ are τ-continuous. Suppose $F = \{f \in C(S) \mid R_S f$ is relatively τ-compact$\}$ is closed under (pointwise) multiplication. Then F is extremely left amenable if and only if every τ-flow (X,S) has a fixed point.

2.6. Remark: Theorem 2.5 may be generalized by replacing the single topology τ by a family of such topologies. Formulation of this result is left to the reader.

2.7. Theorem: (a) [Mitchell (1970)] $LMC(S)$ is extremely left amenable if and only if every separately continuous flow (X,S) has a fixed point.

(b) [Mitchell (1970)] $LUC(S)$ is extremely left amenable if and only if every jointly continuous flow (X,S) has a fixed point.

(c) [Junghenn (1975)] $K(S)$ is extremely left amenable if and only if every separately continuous flow (X,S) which is equicontinuous on compacta has a fixed point.

(d) [Junghenn (1975)] $WAP(S)$ is extremely left amenable if and only if every separately continuous quasi-equicontinuous flow (X,S) has a fixed point.

(e) [Lau (1973)] $AP(S)$ is extremely left amenable if and only if every separately continuous equicontinuous flow (X,S) has a fixed point.

(f) $LWP(S)$ is extremely left amenable if and only if every jointly continuous quasi-equicontinuous flow (X,S) has a fixed point.

(g) $KWP(S)$ is extremely left amenable if and only if every separately continuous quasi-equicontinuous flow (X,S) which is equicontinuous on compacta has a fixed point.

2.8. Remarks: (a) Granirer (1965) has shown that for discrete semigroups S, S is extremely left amenable (i.e., $LUC(S) = B(S)$ is extremely left amenable) if, for each $s, t \in S$, there exists $r \in S$ such that $sr = tr$. Mitchell (1966) proved that the converse is true.

(b) Granirer and Lau (1971) have shown that the finite subgroups of order n are the only subsemigroups S of a locally compact group G which have the property that $LUC(S)$ has a LIM of the form $\sum_{i=1}^{n} a_i\mu_i$, where the μ_i are distinct members of $MM(LUC(S))$,

$$a_i > 0,\ 1 \leq i \leq n,\ \sum_{i=1}^{n} a_i = 1.$$

In particular, if $LUC(G)$ is extremely left amenable, then G consists only of an identity. (This latter result is also true for totally bounded topological groups G and, more generally, for any topological (or semitopological) group that admits a non-trivial continuous homomorphism into a locally compact group [Granirer (1967)].) See Remarks 1.10 in this regard.

(c) Let S be a semitopological semigroup which is algebraically a group, and suppose $AP(S)$ contains a non-constant function. Then $AP(S)$ cannot be extremely left amenable. (If it were, any AP-compactification of S, a non-trivial group, would contain a right zero.) An analogous remark holds for $SAP(S)$ and general semigroups S.

2.9. <u>Lemma</u> [Granirer and Lau (1971)]: Let F be a translation invariant, left m-introverted C*-subalgebra of $C(S)$ containing the constant functions. Then F is extremely left amenable if and only if, for each $f \in F$ and $t \in S$, there exists $\mu \in MM(F)$ such that $T_\mu(f - L_t f) = 0$.

2.10. <u>Theorem</u> [Granirer and Lau (1971)]: Let p denote the topology on $C(S)$ of pointwise convergence, and let F be a translation invariant, left m-introverted C*-subalgebra of $C(S)$ containing the constant functions. Then the following are equivalent:

(a) F is extremely left amenable.

(b) The p-closure of $R_S f$ contains a constant function for each $f \in F$.

(c) The p-closure of $R_S(f - L_t f)$ contains 0 for each $t \in S$, $f \in F$.

Moreover, if F is extremely left amenable, then for each $f \in F$, $\{\mu(f) \mid \mu \in MLIM(F)\} = \{a \mid a1$ is a constant function in the p-closure of $R_S f\}$.

2.11. Remarks: (a) The analog of 1.13 (a) holds here.

(b) If S is discrete and $F = B(S)$, then Theorem 2.10 reduces to Theorem 1 [Granirer (1967)].

CHAPTER V

EXAMPLES

We remind the reader that both [Hofmann and Mostert (1966)] and [Berglund and Hofmann (1967)] contain chapters entitled Examples.

1. STRUCTURE EXAMPLES

1.1. [Ruppert (1973)]: Let X be a compact Hausdorff space with non-closed subset $D \subset X$. An associative multiplication is defined by $xy = y$, $y \in D$, $xy = x$, $y \in X \setminus D$, which makes X a compact right topological semigroup with

$$\Lambda(X) = \{x \in X \mid \lambda_x : X \to X \text{ is continuous}\} = \emptyset$$

and equicontinuous right translations $\{\rho_x \mid x \in X\}$. The only non-trivial right ideal is D, which is not closed and is also the minimal ideal. If, for example, $X = [0,1]$, which is connected, the associated (by Theorem II.2.12) compact topological semigroup Γ can be disconnected.

1.2. [Ruppert (1973)]: Let G and G_1 be compact topological groups which are algebraically but not topologically isomorphic. (The existence of such groups follows from material in 1.5 ahead.)

Let $\alpha: G \to G_1$ be a discontinuous algebraic isomorphism. In

$$S = G \cup G_1 = G \cup \alpha(G)$$

we define multiplication $\circ$ as follows:

$$\left.\begin{aligned} s \circ t &= st \\ \alpha(s) \circ \alpha(t) &= \alpha(s)\alpha(t) \\ \alpha(s) \circ t &= \alpha(s)\alpha(t) \\ s \circ \alpha(t) &= st \end{aligned}\right\} \text{ for all } s, t \in G.$$

This makes S a compact right topological left-group with maximal groups, i.e., minimal right ideals, G and $\alpha(G) = G_1$ that are compact and algebraically but not topologically isomorphic. (Here $\{\rho_x \mid x \in S\}$ is equicontinuous and $\Lambda(S) = \emptyset$.)

1.3. A compact right topological left-group can have maximal groups, i.e., minimal right ideals, that are not closed. Let G be a compact group (topological or right topological) with associated discrete group G_d and algebraic isomorphism $\alpha\colon\ G \to G_d$. The multiplication $\circ$ as in example 2 makes

$$S = G \cup G_d = G \cup \alpha(G)$$

a semigroup. A topology for which S is a compact right topological left-group has as neighbourhood bases $\{\{\alpha(s)\}\}$ for $\alpha(s) \in \alpha(G)$, and for $s \in G$ with open neighbourhood base $\{V^s_\gamma \mid \gamma \in I\}$ <u>in G</u>, sets $V \subset S$ satisfying $V \cap G = V^s_\gamma$ for some γ and $\{\alpha(s) \mid s \in V \cap G\}\backslash V \cap G_d$ is finite. Again $\{\rho_s \mid s \in S\}$ is equicontinuous, but now $\Lambda(S) = G \subset S$. We note that, while there are two minimal right ideals here, there is only one minimal closed right ideal.

1.4. [Ruppert (1974)]: Let $X = [0,2]$ with the usual metric topology and multiplication $\circ$ given by

$$x \circ y = \begin{cases} xy & x, y \in [0,1] \\ 2 - x(2 - y) & x \in [0,1],\ y \in (1,2] \\ y(2 - x) & x \in (1,2],\ y \in [0,1] \\ 2 - (2 - x)(2 - y) & x, y \in (1,2]. \end{cases}$$

Then X is a compact right topological semigroup with $\{\rho_x \mid x \in X\}$ equicontinuous and $\Lambda(S) = 1$. And Γ (as in Theorem II.2.12) is disconnected.

1.5. One kind of compact right topological group [Ruppert (1973, 1975)]: (a) Let G be a compact topological group with a discontinuous automorphism ψ that generates a finite group F' of automorphisms of G. (For example, R/Z is algebraically isomorphic, via an argument using a Hamel basis for R over Q to the weak direct product

$$H' \times Q_1 \times Q_2 \times \dots \times Q_n \times \prod_{\alpha\in J} Q_\alpha,$$

where $Q_i = Q_\alpha = Q$, the rational numbers, H' is the additive group Q (mod 1) and the cardinality of J is the cardinality of the continuum. Then any non-trivial permutation σ of the numbers 1, 2, ..., n, gives an automorphism ψ,

$$\psi(t,q_1,q_2,\dots,q_n,z) = (t,q_{\sigma(1)},q_{\sigma(2)},\dots,q_{\sigma(n)},z),$$

of R/Z, which is discontinuous since $\{(t,0,0,0,\dots) \mid t \in T\}$ is dense in R/Z and left fixed by ψ.) Define a semidirect product multiplication in $G \times F'$ by

$$(s,\psi)(s',\psi') = (s\psi(s'),\psi\psi').$$

Then $G \times F'$ is a compact right topological group with $\{\rho_x \mid x \in G \times F'\}$ equicontinuous and $(s,\phi) \in \Lambda(G \times F')$ if and only if $\phi \in F'$ is a continuous automorphism.

(b) Suppose the automorphism ψ of R/Z described in (a) arises from the cyclic permutation

$$\sigma: 1 \to 2 \to 3 \to \dots \to n \to 1;$$

σ thus generates a cyclic group F of order n. Define an automorphism α of $(R/Z)^n$ by

$$\alpha(s_1,s_2,\dots,s_n) = (s_{\sigma(1)},s_{\sigma(2)},\dots,s_{\sigma(n)}) = (s_2,s_3,\dots,s_n,s_1)$$

and let Γ be the semidirect product of $(R/Z)^n$ and F with multiplication

$$(\underline{s},\sigma^k)(\underline{s'},\sigma^m) = (\alpha^m(\underline{s})\underline{s'},\sigma^{k+m})$$

for $\underline{s}$, $\underline{s'} \in (R/Z)^n$, $k, m = 1, 2, \dots, n$. Γ is a compact topological group. Let

$$H = \{(1,s_2,s_3,\dots,s_n,\sigma^0) \mid s_2,s_3,\dots,s_n \in R/Z\} \subset \Gamma$$

$$M = \{(s,\psi(s),\psi^2(s),\dots,\psi^{n-1}(s),\sigma^k) \mid s \in R/Z,\ k = 1,2,\dots,n\} \subset \Gamma.$$

H is a closed subgroup of Γ containing no non-trivial normal subgroup of Γ, $M \cap H = \{e\}$ and, by Kronecker's Theorem, M is dense in Γ.

As in (a) (except for notation) let $R/Z \times F = G$ have multiplication $(s,\sigma^k)(s',\sigma^m) = (s\alpha^k(s'),\sigma^{k+m})$, which makes G a compact right topological group. The map $\gamma\colon G \to \Gamma$,

$$\gamma(s,\sigma^k) = (s,\psi(s),\psi^2(s),\dots,\psi^{n-1}(s),\sigma^k),$$

is an algebraic antiisomorphism of G onto $M \subset \Gamma$; hence G is homeomorphic to Γ/H (Theorem II.3.3). Also $\dim G = 1$, $\dim \Lambda(G) = 1$ and $\dim \Gamma = n$.

1.6. A distal flow that is not equicontinuous is as follows. Let X be the 2-torus represented as

$$\{(z,w) \mid z, w \text{ are complex numbers, } |z| = |w| = 1\}.$$

Let τ be the homeomorphism of X given by $\tau(z,w) = (z,zw)$. The map τ generates a flow (X,Z) in the obvious way:

$$(n,x) \to \tau^n(x): \quad Z \times X \to X.$$

Clearly, for each integer n, $\tau^n(z,w) = (z,z^n w)$, from which it follows immediately that (X,Z) is distal. To see that (X,Z) is not equicontinuous, consider the sequence $\{(z_m,1)\} = \{(\exp(i\pi/m),1)\}$. Then $(z_m,1) \to (1,1)$, while the distance from $\tau^m(z_m,1)$ to $\tau^m(1,1)$ equals 2 for each m. Note that each minimal subset of X is (isomorphic to) a circle or a finite subgroup of a circle and the restriction of Z to each minimal subset is equicontinuous; thus the enveloping semigroup $E(X,Z)$ is a topological group in the topology of pointwise convergence on X (Theorem II.3.8).

1.7. If, in the previous example, one uses the more complicated homeomorphism τ, given by

$$\tau_1(z,w) = (\exp(i\alpha)z,\ zw),$$

where α is a fixed irrational multiple of π, one gets a distal flow that is not equicontinuous and is also minimal, i.e., $\{\tau_1^n(z,w) \mid n \in Z\}$ is dense in X for all $(z,w) \in X$ [Namioka (1972); p. 208]. (The direct way of determining that this flow is minimal is to use the main result of [Hardy and Littlewood (1914)]; another way is to use results in [Furstenberg (1961)].) Thus the enveloping semigroup $E(X,Z)$ is a compact, right topological, non-topological group (Theorem II.3.8).

1.8. Another example of a distal flow that is not equicontinuous is given in [Ellis (1969); pp. 53-55]. Let

$$G = \left\{ \begin{pmatrix} 1 & x & y \\ 0 & 1 & z \\ 0 & 0 & 1 \end{pmatrix} \;\middle|\; x, y, z \in R \right\}$$

and let H be the subgroup of G whose members have x, y, z $\in$ Z. Then the flow (G/H,G) ((s,tH) $\to$ stH) is distal, but not equicontinuous. The reader is referred to [Ellis (1969)] for the details.

1.9. (This example is due to J. W. Baker.) Let T be the set of regular Borel measures μ on the unit interval [0,1] satisfying $\|\mu\| \leq 1$, $\mu \geq 0$. T is a compact, affine, right topological, non-semitopological semigroup with the weak * topology $\sigma(T,C([0,1]))$ and multiplication $\mu \circ \nu = \|\nu_c\|\mu$, where ν_c is the continuous part of ν. Let

$$S = \{e(x) \mid x \in [0,1]\} \cup \{0\}.$$

Then S is a compact, topological subsemigroup of T, all products in S being equal to zero, and $\overline{\text{co}}S = T$. Note that $\Lambda(T) = \{0\}$; see Theorem II.4.6.

1.10. Let S be the union of three copies of the circle group T; write $S = T_1 \cup T_2 \cup T = \alpha_1(T) \cup \alpha_2(T) \cup T$, where T_i is a copy of T and $\alpha_i\colon \; T \to T_i$ is the isomorphism, i = 1, 2.

Define multiplication in S as follows:

$$\alpha_i(s)\alpha_j(t) = \alpha_i(s)t = t\alpha_i(s) = \alpha_i(s,t),$$

for i, j = 1, 2, and s, t $\in$ T, where st is the usual product in T and T is isomorphically (as a group) embedded in S.

Define a topology on S as follows: if $t = e^{i\theta} \in T$, sets for a neighbourhood basis of

(i) $\alpha_1(e^{i\theta})$ are of the form $\{\alpha_1(e^{i\psi}) \mid \theta \leq \psi < \theta + \delta\} \cup$

$\{\alpha_2(e^{i\psi}) \mid \theta < \psi < \theta + \delta\} \cup \{e^{i\psi} \mid \theta < \psi < \theta + \delta\}$,

where $\delta > 0$.

(ii) $\alpha_2(e^{i\theta})$ are of the form $\{\alpha_2(e^{i\psi}) \mid \theta - \delta < \psi \leq \theta\} \cup$

$\{\alpha_1(e^{i\psi}) \mid \theta - \delta < \psi < \theta\} \cup \{e^{i\psi} \mid \theta - \delta < \psi < \theta\}$,

where $\delta > 0$.

(iii) $e^{i\theta}$ are of the form $\{e^{i\theta}\}$.

Then S is a compact right topological semigroup with T dense in S and $T = \Lambda = \{s \in S \mid \lambda_s \text{ is continuous}\}$. $\alpha_1(T) \cup \alpha_2(T) = K(S)$ and is a minimal left ideal consisting of two algebraically, but not canonically topologically isomorphic maximal subgroups; i.e., the canonical isomorphism between $\alpha_1(T)$ and $\alpha_2(T)$ given by $\alpha_1(t) \to \alpha_2(t)$ is not topological, although $\alpha_1(t) \to \alpha_2(t^{-1})$ is a topological isomorphism. K(S) is compact, but neither $\alpha_1(T)$ nor $\alpha_2(T)$ is compact. Also $\alpha_i(T)$ is not a topological group, $i = 1, 2$.

By mapping an infinite discrete group G into a product of semigroups isomorphic to S, it can be shown [Baker and Milnes (1977)] that $\beta G \simeq PG$ has an infinite number of minimal right ideals. These minimal right ideals are not closed. Also, the maximal subgroups of $K(\beta G)$ are not closed and need not be pairwise canonically topologically isomorphic.

1.11. (The main idea for this example is due to J. W. Baker.) Let S be a locally compact semitopological semigroup such that

$$\lim_{s\to\infty} ss' = \infty, \quad s' \in S$$

(i.e., whenever $s' \in S$ and $\{s_\alpha\}$ is a net that is ultimately outside every compact $K \subset S$, then $\{s_\alpha s'\}$ is another such net). The following construction shows that S has no maximal right topological compactification (ψ, X) (Definition III.2.3); for, the following construction shows X would have to have arbitrarily large cardinality.

Let $\mathcal{O}$ be a fixed limit ordinal, and let

$$T = \{t \mid t \text{ is an ordinal} < \mathcal{O}\}.$$

We make $S' = (T \times S) \cup \{0\}$ a semigroup by

$$(0,s)(t',s') = (t',ss'),$$

$$(t,s)(t',s') = \begin{Bmatrix} (t' \circ t,\ s), \text{ if } t' \circ t < \mathcal{O} \\ 0, \text{ otherwise} \end{Bmatrix}, \text{ if } t \neq 0,$$

$$(t,s)0 = 0(t,s) = 00 = 0,$$

where $t' \circ t$ is the ordinal sum of t' and t (i.e., the well-ordered set t followed by the well-ordered set t'). We topologize S' by singling out a fixed element $s_0 \in S$ and defining basic neighbourhoods of (t,s) to be sets of the form

(i) (t,V), where V is a neighbourhood of s in S, if $s \neq s_0$ or if $(t,s) = (0,s_0)$.

(ii) $(t,V) \cup (t',W)$, where V is as in (i) and W is the complement of a compact subset of S, if $s = s_0$ and $t = t' + 1$ is not a limit ordinal.

(iii) $(t,V) \cup (X \times S)$, where V is as in (i) and

$$X = \{t' \mid t'' < t' < t \text{ for some } t'' < t\},$$

if $s = s_0$ and t is a limit ordinal.

Basic neighbourhoods of 0 are of the form $\{0\} \cup (Y \times S)$, where $Y = \{t' \mid t'' < t' < 0\}$ for some $t'' < 0$.

Then S' is a compact right topological semigroup and any closed subsemigroup of S' containing $\{0\} \times S \simeq S$ contains $T \times \{Ss_0\}$. (In particular, the closure of $\{0\} \times S$ is not a subsemigroup of S'.)

2. EXTENSION EXAMPLES AND EXAMPLES TO SHOW THE SUBSPACES CAN BE DIFFERENT

1. Let G be a non-compact locally compact group. Then WG is a compact semitopological non-topological semigroup; hence

$$C(WG) = \mathcal{WAP}(WG) \neq \mathcal{AP}(WG) = \mathcal{LUC}(WG) = \mathcal{LWP}(WG).$$

If T = [0,1] with the left-zero multiplication xy = x for all x, y $\in$ T, then $\mathcal{AP}(T) = C(T)$, but $\mathcal{SAP}(T)$ contains only the constant functions. Let S be the disjoint union $G \cup WG \cup T$ with multiplication $\cdot$ defined by $x \cdot y = xy$ if x and y are both in G, both in WG or both in T and $x \cdot y = y$ otherwise. We then have for S, with all indicated containments proper

$$\begin{array}{ccccccc} \mathcal{SAP} & \subset & \mathcal{AP} & \subset & \mathcal{LWP} & \subset & \mathcal{WAP} \\ & & & & \mathcal{LUC} & \subset & \mathcal{WLUC} \\ & & & & \cap & & \cap \\ & & & & \mathcal{LMC} & \subset & C(S). \end{array}$$

And $\mathcal{WAP} \not\subset \mathcal{LUC}$, $\mathcal{LUC} \not\subset \mathcal{WAP}$.

2. [Berglund and Milnes (1977)]. Let S = [0,1] $\times$ R, where [0,1] and the real numbers R have their usual metric and topology, be given the left-group multiplication $(x,s)(y,t) = (x,s+t)$, which makes S a topological semigroup. A function $f \in \mathcal{WAP} \setminus \mathcal{LUC}$ is defined by

$$f(x,s) = \sum_n h_n(x) f_n(s), \text{ where}$$

$$h_n(x) = \begin{cases} 2^{n+2}x - 2, & 2^{-n-1} \leq x \leq 3(2^{-n-2}) \\ -2^{n+2}x + 4, & 3(2^{-n-2}) \leq x \leq 2^{-n} \\ 0, & \text{otherwise,} \end{cases}$$

$$f_n(s) = \begin{cases} (s - n)(n + 1 - s), & n \leq s \leq n + 1 \\ 0, & \text{otherwise.} \end{cases}$$

Thus, although left-groups are perhaps the semigroups closest to being groups, the inclusion diagram of III.14.11 has already been disrupted: $S = [0,1] \times R$ has the same inclusion diagram as in Example 2.1 (all indicated inclusions being proper). One does have that $LUC(S)$ is precisely the subset of $C(S)$ whose members are uniformly continuous with respect to the natural product uniformity on $[0,1] \times R$.

3. (a) If $S = R \cup \{\theta\}$ is the one-point compactification of R with multiplication (which is addition in R) extended from R to S by defining $\theta s = s\theta = \theta\theta = \theta$, then S is a compact semitopological semigroup and the only functions in $SAP(R) = AP(R)$ that extend to functions in $C(S)$ are the constant functions. (Also, the only functions in $LUC(R) = LMC(R)$ that extend to functions in $C(S)$ are those which have a limit as $|s| \to \infty$.)

(b) Adjoin a zero θ at $+\infty$ to R to get a (non-compact) topological semigroup $S = R \cup \{\theta\}$ where, for a net $\{s_\alpha\} \subset R$,

$s_\alpha \to \theta$ is equivalent to $s_\alpha \to +\infty$. Again, the only functions in $SAP(R) = AP(R)$ that extend to functions in $C(S)$ are the constant functions. Also of interest here is the function $s \to \tan^{-1}s$, which is in $LUC(R) = LMC(R)$ and has an extension in $C(S)$; but the extension is not in $LMC(S)$. For, the sequence of translates of this extension by the members $\{-n\} \subset R$ converges to a function f with $f(s) = -\frac{\pi}{2}$ for $s \in R$, $f(\theta) = \frac{\pi}{2}$: f is not continuous and the extension is not in $LMC(S)$.

We remark that these examples are best possible in the sense that one cannot get a compact topological semigroup S with dense subgroup G such that a function in $AP(G)$ (or in $LMC(G)$, for that matter) fails to extend to a function in $AP(S)$; for, the hypotheses here imply that S is a group [Hewitt and Ross (1970); 27.39], and Corollary III.15.6 applies.

4. Consider the subsemigroup $(0,\infty)$ of R. Then $s \to \sin(1/s)$ is in $LUC(0,\infty)$, but has no continuous extension to R.

5. [Milnes and Pym (1977)]. Let H be defined for $t \in [0,1]$ by $H(t) = t(1 - t)$. (Any non-constant function having the value zero at the endpoints would do.) A function $h \in C(R) \setminus LUC(R)$ is defined by

$$h(t) = H(t(\text{mod } 1))H(t(\text{mod } 1/n!))$$

for $t \in [n,n+1]$, $n = 0, 1, 2, \ldots$, $h(t) = 0$ for $t \leq 0$. However, the restriction of h to the rational numbers Q is in $LMC(Q)$ and so $LMC(Q) \neq LUC(Q)$.

6. [Milnes (1975)]. Let G be the group of finite permutations of a countable infinite set X: each member of G moves only a finite number of members of X. It has been noted by W. A. Veech that the only functions in $AP(G)$ are linear combinations of the constant function 1 and the function that is 1 at even permutations and -1 at odd permutations. Thus, if H is any finite subgroup of G containing 3 or more elements, then not all functions in $AP(H)$ extend to functions in $AP(G)$.

7. [Burckel (1970)]. Let S be the usual non-negative additive real numbers, and let f be a function with compact support in $C(S)$. Then $f \in AP(S)$ and obviously has an extension in $WAP(R)$, but no extension in $AP(R)$.

8. [Milnes (1973)]. Let G be the affine group of the line, the "ax + b" group: $G = \{(a,b) \in R^2 \mid a > 0\}$ with multiplication $(a,b)(a',b') = (aa',ab'+b)$. It follows essentially from the fact that the left and right uniformities on G are not the same that no non-trivial character on the abelian subgroup $H = \{(1,y) \mid y \in R\} \subset G$ extends to a function that is in $AP(G)$ or even in $WAP(G)$. In fact, if f is a non-trivial character on H and $g \in C(G)$ is uniformly continuous with respect to both the left and right uniformities of G, then

$$\|g|_H - f\|_\infty = \sup_{s \in H} |g(s) - f(s)| = 1.$$

Chou (1975) has given a similar example in $T \times C$ with $(t,c)(t',c') = (tt',tc'+c)$.

9. Let G be a non-compact locally compact group which we will regard as canonically embedded (homeomorphically and isomorphically) in WG. Then $\mathcal{WAP}(G) = \mathcal{LWP}(G)$, but $\mathcal{WAP}(WG) = \mathcal{C}(WG) \neq \mathcal{LWP}(WG) = \mathcal{AP}(WG)$ (Remark III.14.5 (ii)). Thus not all functions in $\mathcal{LWP}(G)$ extend to functions in $\mathcal{LWP}(WG)$.

Appendix A

An Approach through Category Theory

1. While we have not taken a category theory approach to the various topologico - algebraic structures examined in these notes, we have, nonetheless, maintained some notation and terminology reminiscent of categories. In the same way as the various compactifications in [Berglund & Hofmann (1967)] are produced through the Adjoint Functor Theorem, the ones produced here could also be derived. We describe the procedure in this appendix.

2. Adjoint Situations: Roughly speaking, we have an "adjoint situation": if we have categories $\mathcal{A}$ and $\mathcal{B}$ and functors $G : \mathcal{A} \to \mathcal{B}$ and $F : \mathcal{B} \to \mathcal{A}$ such that $F \circ G$ and $G \circ F$ are, up to natural isomorphy,[1] the identity functors on $\mathcal{A}$ and $\mathcal{B}$, respectively. The difficulty lies in starting with a functor $G : \mathcal{A} \to \mathcal{B}$ and proceeding to find a "left-adjoint" functor F possessing the desired property.

3. Definition: If $\mathcal{A}$ and $\mathcal{B}$ are categories, G and F are functors and η and ε are natural transformations such that

---

[1]No attempt is made here to define all the terms used from category theory since that would take us too far afield. In general, undefined terms mean what one might suspect they mean. We adhere to the notation of [Herrlich & Strecker (1973)].

(i) $G : A \to B$ and $F : B \to A$

(ii) $\eta : 1_B \to G \circ F$ and $\varepsilon : F \circ G \to 1_A$

and

(iii) the functorial diagrams

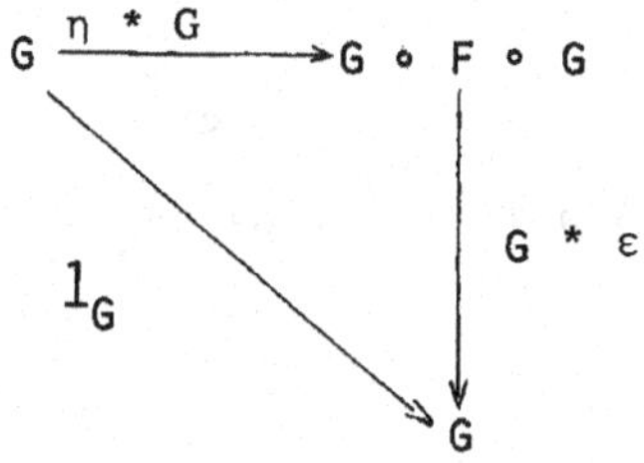

and

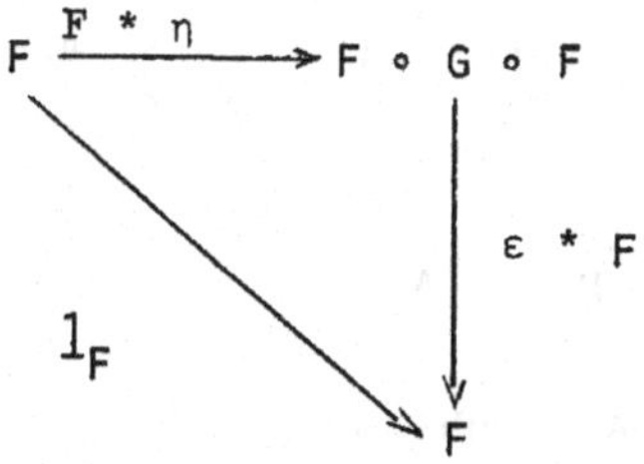

commute,

then this is called an <u>adjoint situation,</u> denoted

$$(\eta, \varepsilon) : F \dashv G.$$

A functor $G : \mathcal{A} \to \mathcal{B}$ is said to <u>have a left adjoint</u> provided that there exist F, η, ε so that an adjoint situation, as above, exists. If each of F and $\hat{F}$ is a left adjoint of the functor G, then F is naturally isomorphic to $\hat{F}$. In an adjoint situation, certain universal properties are observed.

4. <u>Definition</u>: Let $G : \mathcal{A} \to \mathcal{B}$ be a functor, and let B be an object in $\mathcal{B}$. A pair (η_B, A_B), with A_B an object in $\mathcal{A}$, and with

$$\eta_B : B \to G(A_B)$$

a $\mathcal{B}$-morphism, is called a G-universal map for B provided that for each object A' in $\mathcal{A}$ and each $\mathcal{B}$-morphism

$$\phi : B \to G(A')$$

there exists a unique $\mathcal{A}$-morphism

$$\overline{\phi} : A_B \to A'$$

such that the diagram

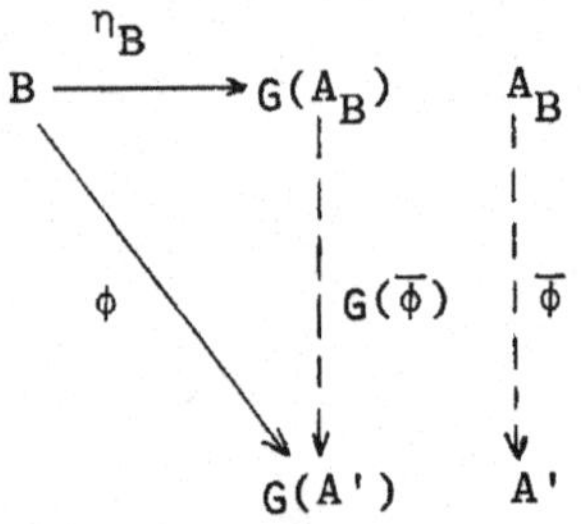

commutes.

If $\mathcal{A}$ is a subcategory of $\mathcal{B}$ and $G : \mathcal{A} \to \mathcal{B}$ is the inclusion functor, if B is an object in $\mathcal{B}$, and (η_B, A_B) is a G-universal map for B, then (η_B, A_B) is called an $\mathcal{A}$-reflection of B.

One often suppresses mention of the morphism η_B when its definition is clear.

5. Theorem [Herrlich & Strecker (1973) p. 196]. Let $G : \mathcal{A} \to \mathcal{B}$ be a functor.

(1) If each object B in $\mathcal{B}$ has a G-universal map (η_B, A_B), then there exists a unique adjoint situation $(\eta, \varepsilon) : F \dashv G$ such that $\eta = (\eta_B)$ and for each object B in $\mathcal{B}$, $F(B) = A_B$.

Conversely,

(2) if $(\eta, \varepsilon) : F \dashv G$ is an adjoint situation, then for each object B in $\mathcal{B}$ the pair $(\eta_B, F(B))$ is a G-universal map for B.

In practice, one often recognizes the existence in nature of an adjoint situation by observing the existence of universal maps. Of particular interest are the various "free" algebraic entities: free groups, free semigroups, free R-modules, etc. Each of these "free" functors is a left adjoint of the forgetful functor into SET.

6. An adjoint situation between two categories induces an

adjoint situation in the dual categories obtained by reversing all arrows:

<u>Theorem</u>: $(\eta,\varepsilon) : F \dashv G$ is an adjoint situation if, and only if, $(\varepsilon,\eta) : G^{op} \dashv F^{op}$ is an adjoint situation.

Thus, if we can identify the dual categories intrinsically, we can move from one adjoint situation to another.

7. <u>Definition</u>: Let $G : A \to B$ be a functor and let B be a B-object. A set-indexed family $\{(u_i, A_i),\ i \in I\}$, where each A_i is an A-object and each $u_i : B \to G(A_i)$ is a B-morphism is called a G-<u>solution set for</u> B provided that for each A-object $\hat{A}$ and each B-morphism $\phi : B \to G(\hat{A})$, there is some $i \in I$ and some $\hat{\phi} : A_i \to \hat{A}$ such that the diagram

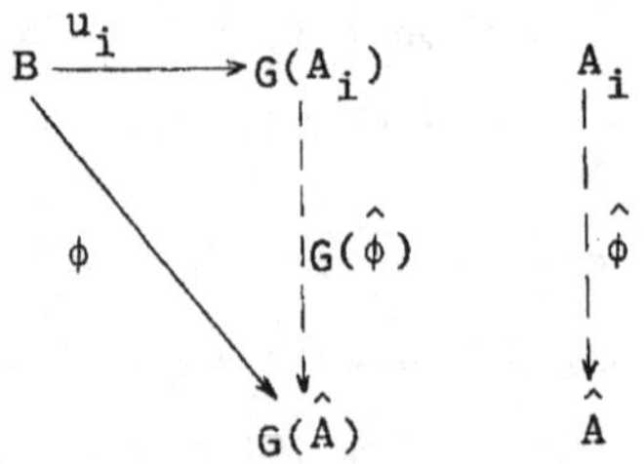

commutes.

<u>Note</u> that the operative idea in the definition is that I is a *set*.

8. Adjoint Functor Theorem: Let A be a complete category, and suppose

$$G : \mathcal{A} \to \mathcal{B}$$

is a functor. Then G has a left adjoint if, and only if,

(1) G preserves limits,

and

(2) each $\mathcal{B}$-object B has a G-solution set.

The criteria for having a *complete* category appear in the proof of 10 below.

9. Definition: Let S be a semitopological semigroup with identity 1, and let T be a right topological (respectively, a semitopological) <respectively, a topological > monoid. Suppose that S acts on T on the left; that is, suppose there is a function $m : S \times T \to T$ with the following properties (where we write $s \cdot t$ for $m(s,t)$):

(1) $(s_1 s_2) \cdot t = s_1 \cdot (s_2 \cdot t)$, for all $s_1, s_2 \in S$ and $t \in T$;

(2) $s \cdot (t_1 t_2) = (s \cdot t_1)t_2$, for all $s \in S$ and all $t_1, t_2 \in T$; and

(3) $1 \cdot t = t$ for all $t \in T$.

In addition, suppose that

(4) $(S \cdot 1)^-$ is a retract of T .

If m is continuous, then T is called a right topological (respectively, a semitopological)

<respectively, a *topological*> S-*module*. If m is separately continuous, then T is called a *separate* right topological (resp., semitopological) < resp., topological> S-module. If m is separately continuous and jointly continuous on compact subsets of $S \times T$, then T is called a right topological (resp., semitopological) < resp., topological> S-*kmodule*

Suppose that $\alpha : T_1 \to T_2$ is a morphism in the category of right topological (resp., semitopological) <resp., topological> monoids. If T_1 and T_2 are S-modules, and if the diagram

$$\begin{array}{ccc} S \times T_1 & \xrightarrow{m_1} & T_1 \\ {\scriptstyle 1_S \times \alpha}\downarrow & & \downarrow{\scriptstyle \alpha} \\ S \times T_2 & \xrightarrow{m_2} & T_2 \end{array}$$

commutes, then α is an S-*morphism*. The condition on α is that

$$\alpha(m_1(s,t)) = \alpha(s \cdot t)$$

$$= s \cdot \alpha(t) = m_2(s,\alpha(t)) .$$

With the above definition of S-morphism, we proceed to define the following categories in the obvious way:

RT S-MOD, is the category of right topological S-modules;

ST S-MOD, the category of semitopological S-modules;

TOP S-MOD, the category of topological S-modules;

RT S-kMOD, the category of right topological S-$k$modules;

ST S-kMOD, the category of semitopological S-$k$modules;

TOP S-kMOD, the category of topological S-$k$modules;

SEP RT S-MOD, the category of separate right topological S-modules;

SEP ST S-MOD, the category of separate semitopological S-modules; and

SEP TOP S-MOD, the category of separate topological S-modules.

Denote the categories of semitopological semigroups by ST SGP , and of topological semigroups by TOP SGP . Also, denote the categories of semitopological and topological groups by ST GRP and TOP GRP, respectively.

Remarks: An object T in SEP RT S-MOD is, in the language of Chapter II, a right topological semigroup with identity in which $\Lambda = \Lambda(S) \neq \emptyset$ and for which Λ^- is a retract (cf., II.2.8). We are really interested in the case where $S \cdot 1$ is dense in T , but use the retract notion to achieve completeness of the category.

10. Theorem All of the thirteen categories named above are complete.

Proof: The proof for SEP RT S-MOD , the category whose objects have the least structure, is presented here. The proofs for the other categories are similar.

A category is complete if it has products and equalizers [Herrlich & Strecker (1973) theorem 23.8, p. 159]. The usual cartesian product with coordinatewise multiplication and distribution of action is a product in the sense of categories [*cf.*, Herrlich & Strecker (1973) p. 117] . $(S \cdot 1)^-$, embedded on the diagonal, is a retract of the product.

Suppose that T_1 and T_2 are objects in SEP RT S-MOD and that

$$T_1 \underset{\beta}{\overset{\alpha}{\rightrightarrows}} T_2$$

are morphisms. We must show that there is an object E in SEP RT S-MOD and a morphism

$$\varepsilon : E \to T_1$$

in SEP RT S-MOD such that

$$\alpha \circ \varepsilon = \beta \circ \varepsilon$$

and such that for any morphism

$$\varepsilon' : E' \to T_1$$

in Sep RT S-Mod with

$$\alpha \circ \varepsilon' = \beta \circ \varepsilon',$$

there is a unique morphism

$$\bar{\varepsilon} : E' \to E$$

such that the diagram

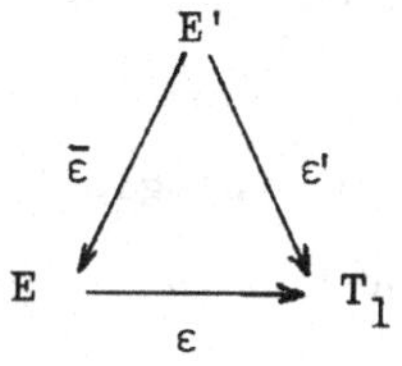

commutes.

The usual construction of an equalizer works in Sep RT S-Mod ; namely, let

$$E = \{t \in T_1 \mid \alpha(t) = \beta(t)\},$$

and

$$\varepsilon : E \to T_1 .$$

be the inclusion. Since, for $s \in S$ and $t \in E$, we have

$$\begin{aligned}\alpha(s \cdot t) &= s \cdot \alpha(t) \\ &= s \cdot \beta(t) \\ &= \beta(s \cdot t) ,\end{aligned}$$

we have that ε is an S-morphism. Since α and β are monoid morphisms, we have that $1 \in E$, hence $S \cdot 1 \subseteq E$. Since E is closed, we get that $(S \cdot 1)^- \subseteq E$. Using the restriction to E of the retraction $T_1 \to (S \cdot 1)^-$, we get that E is in SEP RT S-MOD. That finishes the proof because (E,ε) is clearly an equalizer in the category of right topological semigroups.

11. Definition: Let S be a semitopological monoid. Suppose S acts on an affine monoid T so that conditions (1) - (3) of Definition 9 are satisfied. Suppose also that the following conditions hold:

(4') the closed convex hull $\overline{co}(S \cdot 1)$ is a retract of T, and

(5) the action

$$m : S \times T \to T$$

is an affine function
[that is, for $r \in [0,1]$, we have

$$\begin{aligned} m(s, rt_1 + (1-r)t_2) &= s \cdot (rt_1 + (1-r)t_2) \\ &= r(s \cdot t_1) + (1-r)(s \cdot t_2) \\ &= rm(s,t_1) + (1-r)m(s,t_2)] .\end{aligned}$$

Then T is an <u>affine</u> <u>S-module</u>.

Let T_1 and T_2 be affine S-module which are also separate right topological S-modules. Suppose that $\alpha : T_1 \to T_2$ is a morphism in SEP RT S-MOD such that for $r \in [0,1]$ and $t_1, t_2 \in T_1$,

$$\alpha(rt_1 + (1-r)t_2) = r\alpha(t_1) + (1-r)\alpha(t_2);$$

that is, α is an affine function.

Let A SEP RT S-MOD be the subcategory of SEP RT S-MOD consisting of all such objects and morphisms. [Note that A SEP RT S-MOD is not a *full* subcategory of SEP RT S-MOD].

Analogously define the categories

A SEP ST S-MOD,	A SEP TOP S-MOD,
A RT S-*k*MOD	A ST S-*k* MOD
A TOP S-*k*MOD	A RT S-MOD
A ST S-MOD	A TOP S-MOD
A TOP SGP	A ST SGP
A ST GRP	A TOP GRP

12. <u>Theorem</u>: The thirteen categories of affine things defined above are also complete.

Proof: The proof of Theorem 10 will carry over after observing that the cartesian product of two convex sets is convex and that if α and β are affine morphisms, then

$$\{t \in T_1 \mid \alpha(t) = \beta(t)\}$$

is convex, and the inclusion is an affine morphism.

13. Notation: For any of the twenty-six categories already named, prefix $\mathcal{C}$ to denote the full complete subcategory [Herrlich & Strecker (1973) p. 158] of compact objects.

14. Theorem: Let ONE CAT denote one of the fifty-two categories named in 9, 11, or 13. Let T be an object in ONE CAT. If T is an S-module, let $\check{T}$ denote $(S \cdot 1)^-$ otherwise let $\check{T} = T$. There is a *set*

$$M = \{(\alpha_i, T_i) \mid i \in I\},$$

where T_i is an object in ONE CAT and

$$\alpha_i : T \to T_i$$

is a morphism in ONE CAT, such that

(1) $\alpha_i(T)$ is dense in T_i, $i \in I$,

and such that

(2) given a morphism

$$\alpha' : T \to T'$$

in ONE CAT, there is an element (α_i, T_i) and there is an isomorphism η in ONE CAT,

$$\eta : T_i \to [\alpha'(\check{T})]^-$$

[closure in T'] , such that α' decomposes according to the diagram

$$\begin{array}{ccc} T & \xrightarrow{\xi' \circ \alpha'} & \check{T}' \\ {\scriptstyle \alpha_i}\downarrow & & \uparrow \cup \\ T_i & \xrightarrow[\eta]{} & [\alpha'(\check{T})]^- , \end{array}$$

where $\xi' : T' \to \check{T}'$ is the retraction or identity.

Proof: The proof is the same as for [Berglund & Hofmann (1967) Lemma III. 1.1, p. 112], but for clarity is reproduced here with appropriate modifications.

(I) There is only a set of images of $\check{T}$ in the category of sets since each such image is given (within isomorphy) by a quotient map of $\check{T}$.

(II) For any set X , there is only a set of topologies on X (a subset of the power set of the power set of X).

(III) For any topological space X , there is only a set of topological spaces Y with $X \subseteq Y$ and $\bar{X} = Y$ (up to homeomorphy) since the cardinality of Y is bounded by the cardinality of the set of filters on X .

Let Γ denote a class of morphisms with domain $\check{T}$ such that

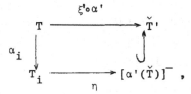

(1) if $\gamma \in \Gamma$, then $\gamma(\check{T})$ is dense in the codomain of γ, and

(2) if $\gamma : \check{T} \to Q$ and $\gamma' : \check{T} \to Q'$ are in Γ, then there is an isomorphism

$$\phi : Q \to Q'$$

such that the diagram

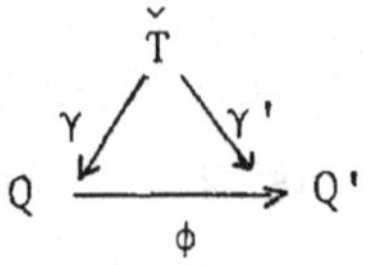

commutes.

From each such class Γ_i, choose one representative γ_i. Let T_i be the codomain of γ_i. By (I),(II), (III), there is only a set of such pairs (γ_i, T_i).

If T is not being considered as a S-module, let $\alpha_i = \gamma_i$ to complete the proof.

If T is an S-module, and $\xi : T \to (S \cdot 1)^-$ is the retraction, let $\alpha_i = \gamma_i \circ \xi$ to complete the proof.

15. <u>The Grand Design</u>: We have the commutative diagram of inclusion functors shown in Figure 1. In addition, if S is a given fixed semitopological semigroup, we have the commutative diagram of inclusion functors shown in Figure 2.

Because of Theorem 14, we may apply the Adjoint Functor Theorem (Theorem 8) to get the commutative diagrams of reflections shown in Figures 3 and 4 . Finally, we may attach the diagram of S-modules to the diagram of semigroups using forgetful functors to produce the commutative diagram of Figure 5: "The Grand Design." (Note that because of the retraction condition on the solution sets for S -modules produced in Theorem 14, the compactifications of Figure 4 are actually compactifications of the retracts.)

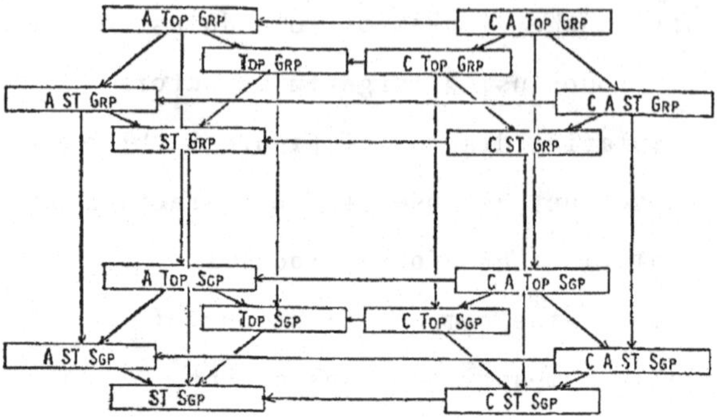

Figure 1

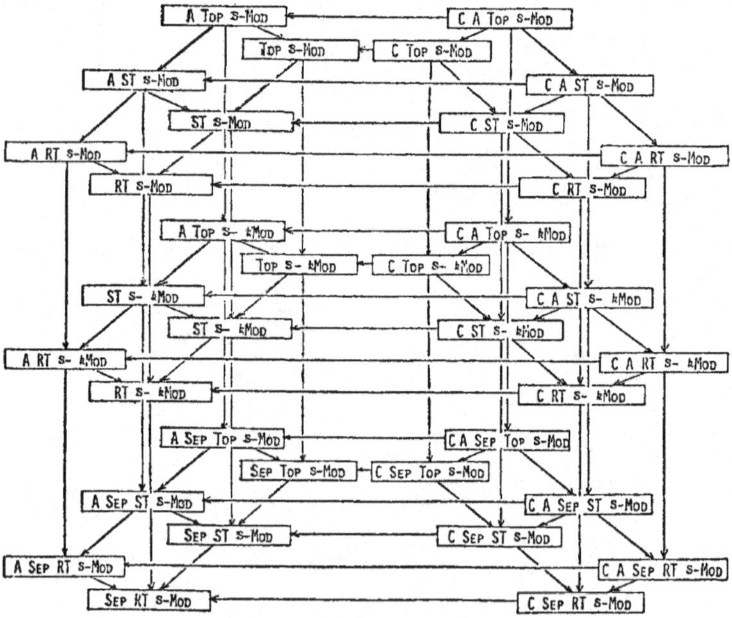

Figure 2

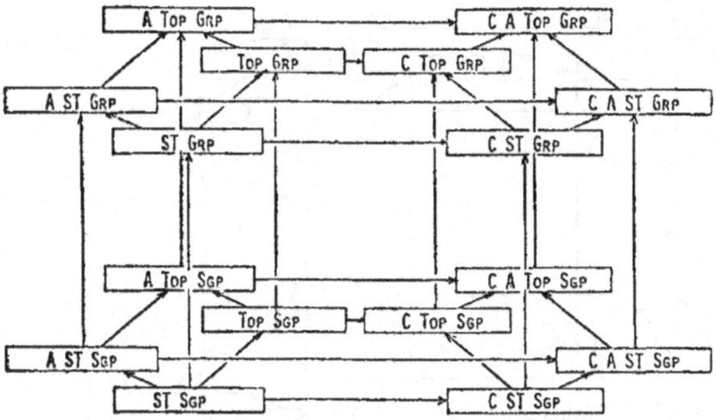

Figure 3

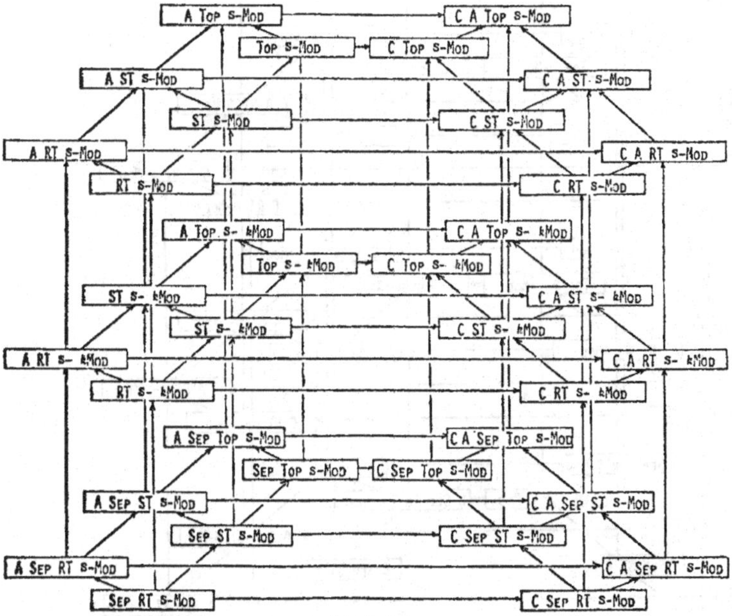

Figure 4

THE GRAND DESIGN

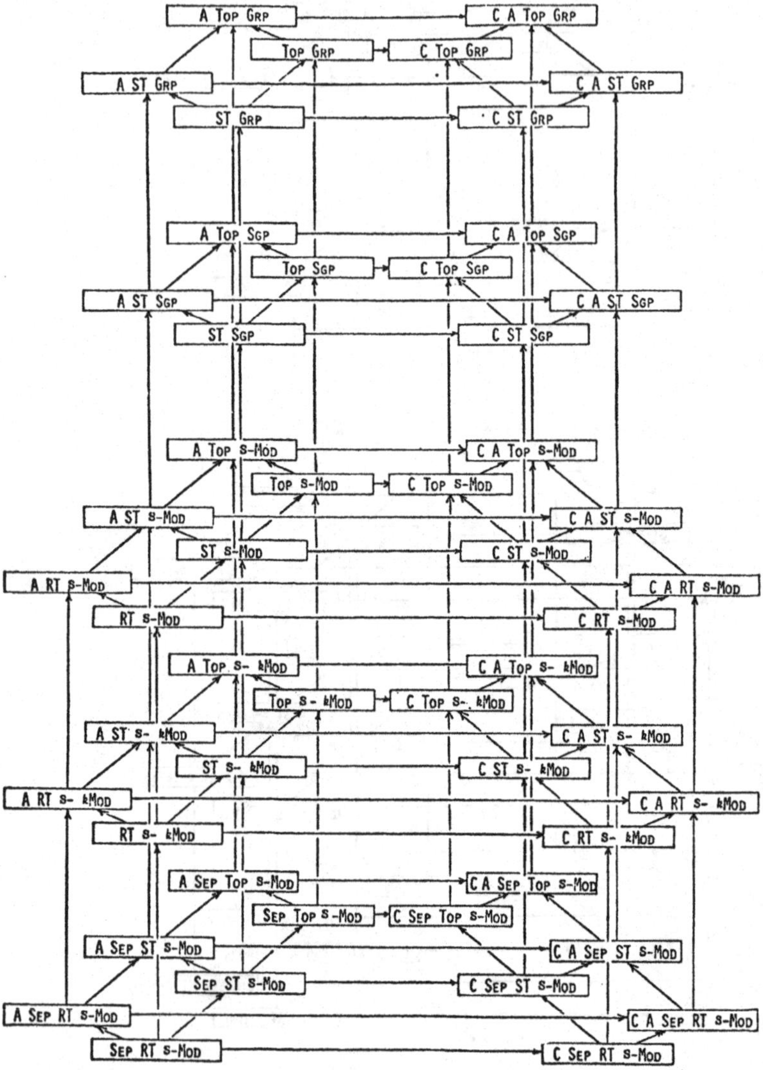

Figure 5

16. Discussion: Culling Figure 5 to reveal the categories of particular interest to us, we get Figure 6. Finally, because we are really only interested in what happens to a given semitopological monoid S considered as a semigroup or as an S-module, we compose several of the functors to obtain Figure 7. (For aesthetic reasons, some of the categories on the left are repeated. The double lines indicate the appropriate identity functor.) Functors of interest are labeled in Figure 7.

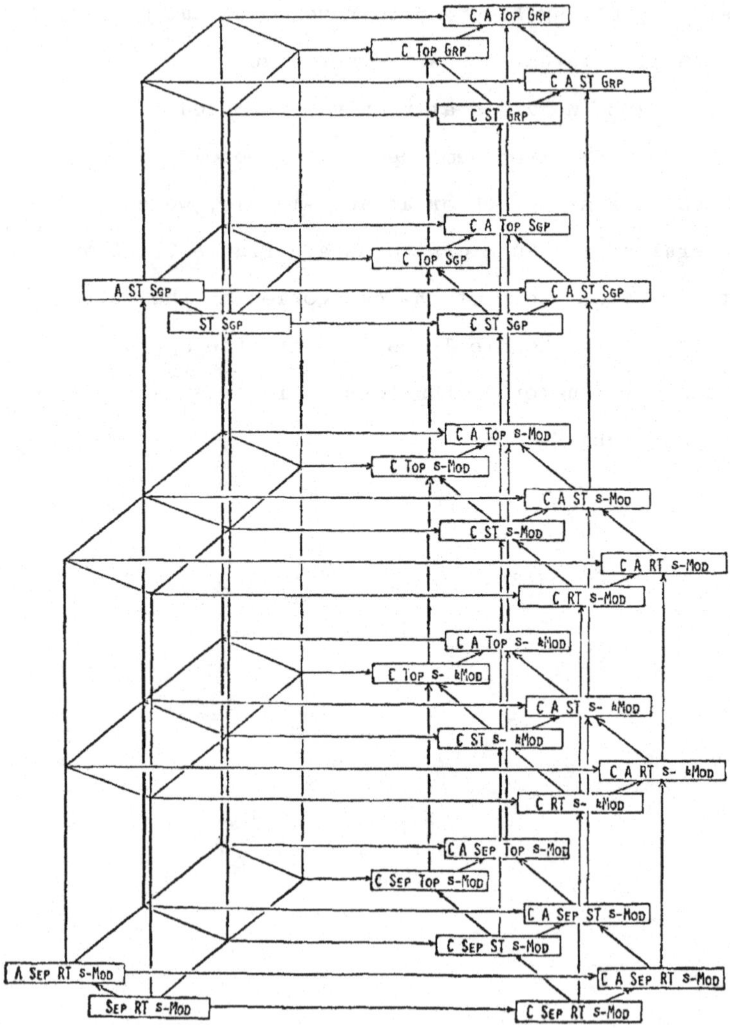
C A Top Grp
C Top Grp
C A ST Grp
C ST Grp
C A Top Sgp
C Top Sgp
A ST Sgp
C A ST Sgp
ST Sgp
C ST Sgp
C A Top s-Mod
C Top s-Mod
C A ST s-Mod
C ST s-Mod
C A RT s-Mod
C RT s-Mod
C A Top s- ℓMod
C Top s- ℓMod
C A ST s- ℓMod
C ST s- ℓMod
C A RT s- ℓMod
C RT s- ℓMod
C A Sep Top s-Mod
C Sep Top s-Mod
C A Sep ST s-Mod
C Sep ST s-Mod
A Sep RT s-Mod
C A Sep RT s-Mod
Sep RT s-Mod
C Sep RT s-Mod

Figure 6

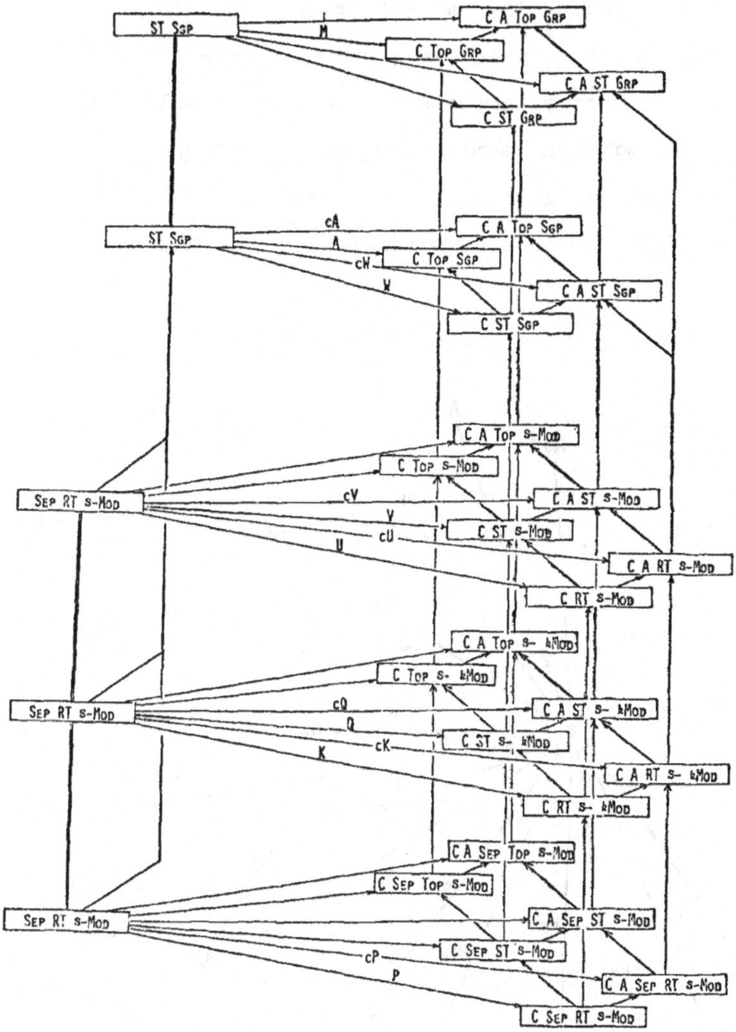
ST Sgp
C A Top Grp
M
C Top Grp
C A ST Grp
C ST Grp
cA
ST Sgp
C A Top Sgp
A
C Top Sgp
cW
C A ST Sgp
W
C ST Sgp
C A Top s-Mod
C Top s-Mod
Sep RT s-Mod
cV
C A ST s-Mod
V
C ST s-Mod
cU
C A RT s-Mod
U
C RT s-Mod
C A Top s- kMod
C Top s- kMod
Sep RT s-Mod
cQ
C A ST s- kMod
Q
C ST s- kMod
cK
C A RT s- kMod
K
C RT s- kMod
C A Sep Top s-Mod
C Sep Top s-Mod
Sep RT s-Mod
C A Sep ST s-Mod
C Sep ST s-Mod
cP
C A Sep RT s-Mod
P
C Sep RT s-Mod

Figure 7

17. Let S be a semitopological semigroup with identity 1. The S-module in which we are most interested is S itself. Identifying the separate right-topological S-module S with the semitopological semigroup S, we get the following collection of "embedding" morphisms from the adjoint situation:

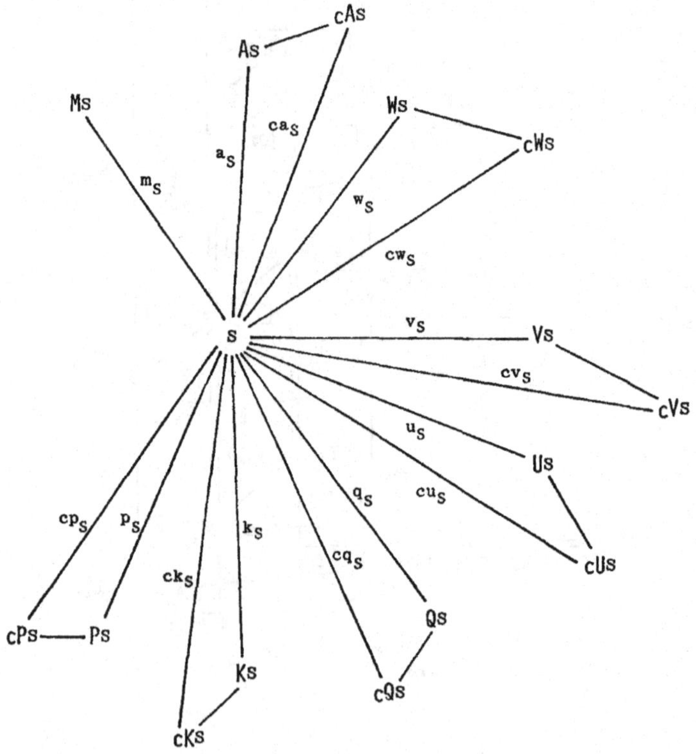

Figure 8

18. Recall from 5 that an adjoint situation gives rise to certain universal properties. Thus, for example, if S and S' are semitopological semigroups, and

$$\phi : S \to S'$$

is a morphism in $\mathcal{ST}\,\mathcal{S}_{GP}$, then there is a unique morphism

$$\phi' : \mathcal{W}S \to \mathcal{W}S'$$

such that the diagram

$$\begin{array}{ccc} \mathcal{W}S & \xrightarrow{\phi'} & \mathcal{W}S' \\ {\scriptstyle w_S}\uparrow & & \uparrow{\scriptstyle w_{S'}} \\ S & \xrightarrow[\phi]{} & S' \end{array}$$

commutes. Similar statements hold for all the other categories.

Although the universal properties hold for modules, some interpretation is required to avoid confusion: Suppose that S and S' are semitopological semigroups with identity, and that

$$\phi : S \to S'$$

is a morphism of semitopological monoids.
Suppose that T' is an S'-module. Then T' is an S-module under the action

$$s \cdot t = \phi(s) \cdot t' , \quad s \in S, \; t' \in T,$$

provided that $\phi(S)^-$ is a retract of S' .

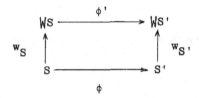

Moreover, if $\alpha : T_1' \to T_2'$ is an S'-morphism, then it is also an S-morphism under the action induced by ϕ. Thus, ϕ induces forgetful functors such as

$$\mathcal{F} = \mathcal{F}_\phi : \text{SEP RT S'-MOD} \to \text{SEP RT S-MOD}.$$

Now suppose that T and T' are objects in SEP RT S-MOD and SEP RT S'-MOD, respectively. Suppose that

$$\psi : T \to \mathcal{F}T'$$

is a morphism in SEP RT S-MOD. Then there is a morphism

$$\psi' = \mathcal{P}\psi' : \mathcal{P}T \to \mathcal{P}\mathcal{F}T'$$

in the category C SEP RT S-MOD such that the diagram

$$\begin{array}{ccc} \mathcal{P}T & \xrightarrow{\psi'} & \mathcal{P}\mathcal{F}T' \\ \uparrow & & \uparrow \\ T & \xrightarrow[\psi]{} & \mathcal{F}T' \end{array}$$

commutes.

In the case that $T = S$, $T' = S'$, and $\psi = \phi$, some confusion arises from the natural inclination to identify S' with $\mathcal{F}S'$. One cannot then readily distinguish between the S'-module $\mathcal{F}S'$ and the S-modules $\mathcal{P}\mathcal{F}S'$ and $\mathcal{F}\mathcal{P}S'$. We shall, nevertheless, make the identification from time to time; usually meaning the S-module $\mathcal{P}\mathcal{F}S'$, but writing $\mathcal{P}S'$.

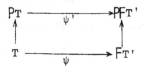

The diagram

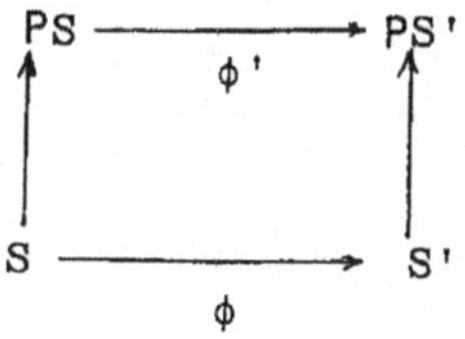

is interpreted to be the commutative diagram of S-modules

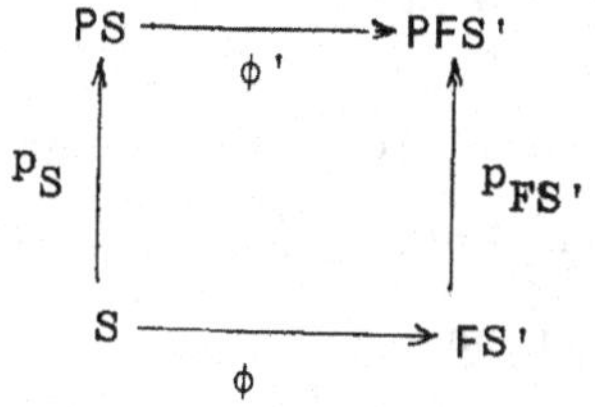

Since FPS' is a compact separate right-topological S-module and $Fp_{S'}$ is a continuous S-morphism, we also get a diagram

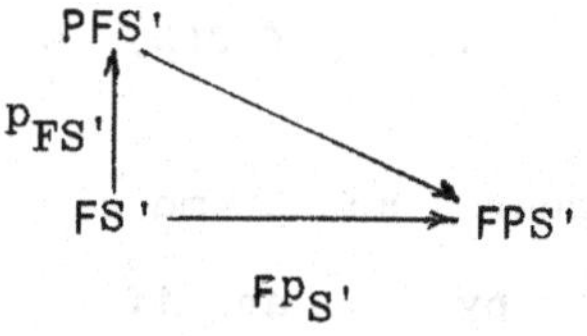

19\. Notation: Let Top_2 denote the category of Hausdorff topological spaces and continuous functions. Let C be the contravariant functor which assigns to each Hausdorff space X the commutative C^*-algebra $C(X)$

of bounded, continuous, complex-valued functions on X and which assigns to each morphism

$$\phi : X \to Y$$

in Top_2 the C*-morphism

$$\phi^* = C(\phi) : C(Y) \to C(X)$$

defined by

$$\phi^* f = f \circ \phi, \ f \in C(Y).$$

Let S be a semitopological semigroup. Suppose that S acts on the Hausdorff space X with a separately continuous action

$$(s,x) \to sx : S \times X \to X.$$

Denote an anti-isomorphic copy of S by S° . The action of S on X induces an action of S° on $C(X)$ given by

$$(s \cdot f)(x) = f(sx), \quad f \in C(X), \ s \in S, \ x \in X.$$

$C(X)$ is, as an algebra, an S°-module. If X and Y are both acted on by S , and if

$$\phi : X \to Y$$

is a continuous function such that

$$\phi(sx) = s \cdot \phi(x), \quad s \in S, \ x \in X,$$

then ϕ^* is an S°-morphism in the sense that

$$\phi^*(s \cdot f) = s \cdot \phi^*(f).$$

Let $\tilde{\phi}$ denote the corestriction of ϕ^* to its image.

20. Definition: Let S be a semitopological semigroup. Define the sub-C*-algebras of $\mathcal{C}(S)$ by

$$\mathcal{M}(S) = \tilde{m}_S[\mathcal{C}(\mathfrak{M}S)],$$

$$\mathcal{A}(S) = \tilde{a}_S[\mathcal{C}(\mathfrak{A}S)],$$

and

$$\mathcal{W}(S) = \tilde{w}_S[\mathcal{C}(\mathfrak{U}S)].$$

Let T be a separate right topological S-module. Define the S°-module sub-C*-algebras of $\mathcal{C}(T)$ by

$$\mathcal{V}(T) = \tilde{v}_T[\mathcal{C}(\mathfrak{V}T)]$$

$$\mathcal{U}(T) = \tilde{u}_T[\mathcal{C}(\mathfrak{U}T)]$$

$$\mathcal{Q}(T) = \tilde{q}_T[\mathcal{C}(\mathfrak{Q}T)]$$

$$\mathcal{K}(T) = \tilde{k}_T[\mathcal{C}(\mathfrak{K}T)]$$

$$\mathcal{P}(T) = \tilde{p}_T[\mathcal{C}(\mathfrak{P}T)].$$

(We are particularly interested in the case where T = S .)

21. Discussion: We wish to establish the following identifications:

$$\mathcal{M}(S) = SAP(S)$$
$$\mathcal{A}(S) = AP(S)$$
$$\mathcal{W}(S) = \mathcal{W}AP(S)$$
$$\mathcal{V}(S) = \mathcal{L}\mathcal{W}P(S)$$
$$\mathcal{U}(S) = \mathcal{L}\mathcal{U}C(S)$$
$$\mathcal{Q}(S) = \mathcal{K}\mathcal{W}P(S)$$
$$\mathcal{K}(S) = K(S)$$
$$\mathcal{P}(S) = \mathcal{L}\mathcal{M}C(S)$$

To do that we need only show that the result of applying the functors

$$\mathsf{M}, \mathsf{A}, \mathsf{W}, \mathsf{V}, \mathsf{U}, \mathsf{Q}, \mathsf{K}, \mathsf{P}$$

to a semitopological semigroup S is to yield precisely the same compactifications

$$MS,\ AS,\ WS,\ VS,\ US,\ QS,\ KS,\ PS$$

constructed earlier.

22. Theorem: Let S_1 and S_2 be a semitopological semigroups, and let T_1 and T_2 be separate right-topological S-modules. Suppose that

$$\alpha : S_1 \to S_2$$

is a morphism of semitopological semigroups, and that

$$\beta : T_1 \to T_2$$

is a morphism of separate right-topological S_1-modules. Then there are unique C*-morphisms

$$M(\alpha) : M(S_2) \to M(S_1),$$

$$A(\alpha) : A(S_2) \to A(S_1),$$

$$W(\alpha) : W(S_2) \to W(S_1),$$

and unique C*-S°-morphisms

$$V(\beta) : V(T_2) \to V(T_1),$$

$$U(\beta) : U(T_2) \to U(T_1),$$

$$Q(\beta) : Q(T_2) \to Q(T_1),$$

$$K(\beta) : K(T_2) \to K(T_1),$$

$$P(\beta) : P(T_2) \to P(T_1).$$

Proof: Similar to [Berglund and Hofmann (1967) III.1. 8, p. 117]

23. Let COMM C* denote the category of commutative C*-algebras with identity. If S° is a semigroup, let COMM C* S°-MOD denote the sub-category of S°-modules and S°-morphisms. From the above, it is clear that we have contravariant functors

$$M, A, W : \text{ST SGP} \to \text{COMM C*}$$

and

$$V, U, Q, K, P : \text{Sep RT } S\text{-Mod} \to \text{Comm } C^*S^\circ\text{-Mod}$$

24. Notation: Let S be a semitopological semigroup, and let T be a separate right topological S-module. Define

$$p_T : T \to L(C(T), C(S))$$

by

$$[p_T(t)f](s) = f(s \cdot t)$$

$$= [C(\rho_t)f](s) \qquad s \in S,\ f \in C(T),\ t \in T\ .$$

25. Let S be a semitopological semigroup. Consider MS, AS, WS, and S to be S-modules. Fix $t \in S$ Then we get the following equations:

$$p_S(t)\tilde{m}_S = \tilde{m}_S p_{MS}(m_S(t))$$

$$p_S(t)\tilde{a}_S = \tilde{a}_S p_{AS}(a_S(t))$$

$$p_S(t)\tilde{w}_S = \tilde{w}_S p_{WS}(w_S(t))$$

$$p_S(t)\tilde{v}_S = \tilde{v}_S p_{VS}(v_S(t))$$

$$p_S(t)\tilde{u}_S = \tilde{u}_S p_{US}(u_S(t))$$

$$p_S(t)\tilde{q}_S = \tilde{q}_S p_{QS}(q_S(t))$$

$$p_S(t)\tilde{k}_S = \tilde{k}_S p_{KS}(k_S(t))$$

$$p_S(t)\tilde{p}_S = \tilde{p}_S p_{PS}(p_S(t)).$$

In particular, $M(S)$, $A(S)$, $W(S)$, $V(S)$, $U(S)$, $Q(S)$, $K(S)$, and $P(S)$ are invariant under the semigroup of operators $p_S(S)$.

26. Notation: Let CNVX denote the subcategory of TOP_2, consisting of convex sets and affine functions. Let

$$F : \mathrm{TOP}_2 \to \mathrm{CNVX}$$

be the left-adjoint of the inclusion functor; and let η be the unit of that adjunction. Let Aff be the contravariant functor which assigns to each Hausdorff topological space X the Banach subspace $Aff(X)$ of $C(X)$ defined by

$$Aff(X) = \{C(\eta_X)f \mid f \in C(FX) \text{ and } f \text{ is affine}\}$$

and which assigns to each continuous function

$$\phi : X \to Y$$

the morphism of Banach spaces

$$Aff(\phi) : Aff(Y) \to Aff(X)$$

defined by

$$Aff(\phi)\, g = g \circ \phi, \; g \in Aff(Y)$$

Let S be a semitopological semigroup. Suppose that S acts on the Hausdorff space X with a separately continuous action

$$(s,x) \to sx : S \times X \to X$$

Denote an anti-isomorphic copy of S by S°. The action of S° on $C(X)$, when restricted to $\mathcal{A}ff(X)$ will induce a vector space action of S on $\mathcal{A}ff(X)$. Thus $\mathcal{A}ff(X)$ is a sub-S°-module Banach subspace of $C(X)$.

27. Definition: For a Banach subspace F containing the constants of some $C(X)$, $\mathcal{M}(F) = \{\mu \in F^* | \mu$ is a mean $\}$, (see I.3.1) and for a bounded linear transformation $\phi: F_1 \to F_2$, where F_1 and F_2 are such Banach spaces, define

$$\tilde{\phi} = \mathcal{M}(\phi) : \mathcal{M}(F_2) \to \mathcal{M}(F_1)$$

by

$$\langle x, \tilde{\phi}\mu \rangle = \langle \phi x, \mu \rangle \quad , x \in F_1, \ \mu \in \mathcal{M}(F_2) .$$

28. Theorem: The corestriction of the contravariant functor

$$\mathcal{A}ff : \mathbf{CCnvx} \to \mathbf{BanSp}$$

is a duality with adjoint (= "inverse") $\mathcal{M}$.

Proof: This result follows from I.3.6 and I.3.7 together with the structure of the duality between $\mathbf{CTop}_2$ and $\mathbf{CommC^*}$.

29. Definition: Let S be a semitopological semigroup. Define Banach subspaces of $A\delta\delta(S)$ by

$$c\mathcal{M}(S) = \tilde{c}m_S\,[A\delta\delta\ (c\mathrm{M}S)],$$

$$c\mathcal{A}(S) = \tilde{c}a_S\,[A\delta\delta\ (c\mathrm{A}S)],$$

$$c\mathcal{W}(s) = \tilde{c}w\ [A\delta\delta\ (c\mathrm{W}S)],$$

Let T be a separate right topological S-module. Define the S°-module Banach subspaces of $A\delta\delta(T)$ by

$$c\mathcal{V}(T) = \tilde{c}v_T\,[A\delta\delta\ (c\mathrm{V}T)]\ ,$$

$$c\mathcal{U}(T) = \tilde{c}u_T\,[A\delta\delta\ (c\mathrm{U}T)]\ ,$$

$$c\mathcal{Q}(T) = \tilde{c}q_T\,[A\delta\delta\ (c\mathrm{Q}T)]\ ,$$

$$c\mathcal{K}(T) = \tilde{c}k_T\,[A\delta\delta\ (c\mathrm{K}T)]\ ,$$

$$c\mathcal{P}(T) = \tilde{c}p_T\,[A\delta\delta\ (c\mathrm{P}T)]\ .$$

30. Theorem: Let S_1 and S_2 be semitopological semigroups, and let T_1 and T_2 be separate right topological S_1-modules.

Suppose that

$$\alpha\ :\ S_1 \to S_2$$

is a morphism of semitopological semigroups, and that

$$\beta : T_1 \to T_2$$

is a morphism of separate right topological S_1-modules. Then there are unique Banach space morphisms (i.e., bounded linear transformations)

$$c\mathcal{M}(\alpha) : c\mathcal{M}(S_2) \to c\mathcal{M}(S_1),$$

$$c\mathcal{A}(\alpha) : c\mathcal{A}(S_2) \to c\mathcal{A}(S_1)$$

$$c\mathcal{W}(\alpha) : c\mathcal{W}(S_2) \to c\mathcal{W}(S_1),$$

and there are unique $S°$-module Banach space morphisms

$$c\mathcal{V}(\beta) : c\mathcal{V}(T_2) \to c\mathcal{V}(T_1),$$

$$c\mathcal{U}(\beta) : c\mathcal{U}(T_2) \to c\mathcal{U}(T_1),$$

$$c\mathcal{Q}(\beta) : c\mathcal{Q}(T_2) \to c\mathcal{Q}(T_1),$$

$$c\mathcal{K}(\beta) : c\mathcal{K}(T_2) \to c\mathcal{K}(T_1), \quad \text{and}$$

$$c\mathcal{P}(\beta) : c\mathcal{P}(T_2) \to c\mathcal{P}(T_1).$$

Proof: The proof is similar to [Berglund and Hofmann (1967) III. 1.8, p. 117].

31. Let BAN SP^1 denote the full subcategory of Banach spaces and bounded linear transformations which are as in Definition 27. If S is a semigroup, let BAN $S°\text{-MOD}^1$ denote the subcategory of $S°$-modules and $S°$ morphisms. From the above, it is clear that we have contravariant functors

$$c\mathcal{M},\ c\mathcal{A},\ c\mathcal{W} : \mathrm{ST\ SGP} \to \mathrm{BAN\ SP}^{1}$$

and

$$c\mathcal{V},\ c\mathcal{U},\ c\mathcal{Q},\ c\mathcal{K},\ c\mathcal{P} : \mathrm{SEP\ RT\ S\text{-}MOD} \to \mathrm{BAN\ S^{\circ}\text{-}MOD}^{1}.$$

<u>Notation</u>: Let S be a semitopological semigroup, and let T be a separate right topological S-module. Define

$$ap_T : T \to L(A\ell\ell\,(T), A\ell\ell\,(S))$$

by

$$[ap_T(t)f\,](s) = f(s \cdot t)$$

$$= [A\ell\ell\,(\rho_t)f\,](s)$$

32. Let S be a semitopological semigroup. Consider $c\mathcal{M}S$, $c\mathcal{A}S$, $c\mathcal{W}S$, and S to be S-modules. Fix $t \in S$. Then we get the following equations:

$$ap_S(t)\tilde{cm}_S = \tilde{cm}_S ap_{c\mathcal{M}S}(cm_S(t)),$$

$$ap_S(t)\tilde{ca}_S = \tilde{ca}_S ap_{c\mathcal{A}S}(ca_S(t)),$$

$$ap_S(t)\tilde{cw}_S = \tilde{cw}_S ap_{c\mathcal{W}S}(cw_S(t)),$$

$$ap_S(t)\tilde{cv}_S = \tilde{cw}_S ap_{c\mathcal{V}S}(cv_S(t)),$$

$$ap_S(t)\tilde{cu}_S = \tilde{cu}_S ap_{c\mathcal{U}S}(cu_S(t)),$$

$$ap_S(t)\tilde{cq}_S = \tilde{cq}_S ap_{c\mathcal{Q}S}(cq_S(t)),$$

$$ap_S(t)\tilde{ck}_S = \tilde{ck}_S ap_{c\mathcal{K}S}(ck_S(t)),$$

$$ap_S(t)\tilde{cp}_S = \tilde{cp}_S ap_{c\mathcal{P}S}(cp_S(t)).$$

In particular, $c\mathcal{M}(S)$, $c\mathcal{A}(S)$, $c\mathcal{W}(S)$, $c\mathcal{V}(S)$, $c\mathcal{U}(S)$, $c\mathcal{Q}(S)$, $c\mathcal{K}(S)$, and $c\mathcal{P}(S)$ are invariant under the semigroup of operators $ap_S(S)$.

33. <u>Theorem</u>: Let S be a semitopological semigroup. The following statements relating $\mathcal{C}$-reflections $(h_S, \mathcal{H}S)$ to $\mathcal{F}$-compactifications (ψ, X) are true:

(i) $(m_S, \mathcal{M}S) = (m, MS)$.

(ii) $(a_S, \mathcal{A}S) = (a, AS)$

(iii) $(w_S, \mathcal{W}S) = (w, WS)$.

Also, considering S to be an S-module, we get the following

(iv) $(v_S, \mathcal{V}S) = (v, VS)$.

(v) $(u_S, \mathcal{U}S) = (u, US)$.

(vi) $(q_S, \mathcal{Q}S) = (q, QS)$

(vii) $(k_S, \mathcal{K}S) = (k, KS)$

(viii) $(p_S, \mathcal{P}S) = (p, PS)$

Proof. The condition in Definition III.2.1 of an $\mathcal{F}$-compactification (ψ, X) that maps

$$\lambda_{\psi(s)} : X \to X$$

be continuous for each $s \in S$ is actually the condition that X be an S-module. The result then follows from the universal properties guaranteed by the adjoint situation (see A.5) combined with the appropriate theorem describing the compactifications: namely Theorems III.10.4, 9.4, 8.4, 11.3, 5.5, 12.3, 6.3, and 4.5, respectively.

34. Corollary: Let S be a semitopological semigroup. The following statements about sub-C*-algebras of $C(S)$ are true:

(i) $\mathcal{M}(S) = SAP(S)$.

(ii) $\mathcal{A}(S) = AP(S)$.

(iii) $\mathcal{W}(S) = WAP(S)$.

Considering S to be an S-module, we also have the following statements about $S°$-module sub-C*-algebras at $C(S)$:

(iv) $\mathcal{V}(S) = LWP(S)$.

(v) $\mathcal{U}(S) = LUC(S)$.

(vi) $\mathcal{Q}(S) = KWP(S)$.

(vii) $\mathcal{K}(S) = K(S)$.

(viii) $\mathcal{P}(S) = LMC(S)$.

35. Theorem: Let S be a semitopological semigroup. The following statements relating $\mathcal{CA}$-universal maps $(ch_S, \mathcal{CH}S)$ to F-affine compactifications are true:

(i) $(cm_S, c\mathcal{M}S) = (cm, cMS)$.

(ii) $(ca_S, c\mathcal{A}S) = (ca, cAS)$.

(iii) $(cw_S, c\mathcal{W}S) = (cw, cWS)$.

Also, considering S to be an S-module, we get the following:

(iv) $(cv_S, c\mathcal{V}S) = (cv, CVS)$.

(v) $(cu_S, c\mathcal{U}S) = (cu, cUS)$.

(vi) $(cq_S, c\mathcal{Q}S) = (cq, cQS)$.

(vii) $(ck_S, c\mathcal{K}S) = (ck, cKS)$.

(viii) $(cp_S, c\mathcal{P}S) = (cp, cPS)$.

Proof: Similar to A.33. Note that $c\mathcal{M}S = cMS$ is the trivial one-point group.

36. <u>Corollary</u>: Let S be a semitopological semigroup. The following statements about sub-Banach spaces of $A\mathfrak{f}\mathfrak{f}(S)$ are true:

(i) $c\mathcal{M}(S) = C$, the complex numbers.

(ii) $c\mathcal{A}(S) = \mathcal{AP}(S)$.

(iii) $c\mathcal{W}(S) = \mathcal{WAP}(S)$.

Considering S to be an S-module we also have the following statements about S°-module Banach subspaces of $A\mathfrak{f}\mathfrak{f}(S)$:

(iv) $c\mathcal{V}(S) = \mathcal{LWP}(S)$.

(v) $c\mathcal{U}(S) = \mathcal{LUC}(S)$.

(vi) $c\mathcal{Q}(S) = \mathcal{CKWP}(S)$.

(vii) $c\mathcal{K}(S) = \mathcal{CK}(S)$

(viii) $c\mathcal{P}(S) = \mathcal{WLUC}(S)$.

37. <u>Example</u>: Let $[0,\infty]$ denote the two-point compactification of the multiplicative group R^+ of positive real numbers. For $a,b \in [0,\infty]$ define

$$ab = \begin{cases} ab & \text{if } a,\ b \in R^+ \\ b & \text{if } b \in \{0,\infty\} \\ a & \text{if } a \in \{0,\infty\}\ \&\ b \in R^+ \end{cases}$$

Then $[0,\infty]$ is a right topological semigroup with identity 1. (In fact, it is an object in SEP RT R^+-MOD.) Note that left translations by 0 and by ∞ are discontinuous.

Let G^o be the group-with-zero which is the one-point compactification of R^+ with the point at infinity z acting as a zero.

Now let

$$T : G^o \times \{0\} \cup [0,\infty] \times \{1\} .$$

Define multiplication in T as follows:

$$(a,1)(b,0) = (b,0)(a,1) = (z,0)$$

if $a \in \{0,\infty\}$; otherwise,

$$(a,e)(b,f) = (ab,ef).$$

Let S be the semitopological subsemigroup of T defined by

$$S = G^o \times \{0\} \cup R^+ \times \{1\} .$$

With the natural action, T is an object in SEP RT S-MOD . Applying the various functors of Figure 7 (and letting

$$F : \mathrm{Sep\ RT\ S\text{-}Mod} \to \mathrm{ST\ Sgp}$$

be the coadjoint of the forgetful functor), we get the following results:

$$p_T : T \to \mathcal{P}T$$

is the identity morphism;

$$u_T : T \to \mathcal{U}T$$

identifies $G^o \times \{0\}$ to a point (an isolated zero);

$$v_T : T \to \mathcal{V}T$$

identifies $G^o \times \{0\}$ to a point and identifies $(0,1)$ and $(\infty,1)$;

$$w_T : \mathcal{F}T \to \mathcal{W}\mathcal{F}T$$

identifies $(0,1)$ and $(\infty,1)$;

$$a_T : \mathcal{F}T \to \mathcal{M}\mathcal{F}T$$

identifies $G^o \times \{0\}$ to a point and $[0,\infty] \times \{1\}$ to a point;

$$m_T : \mathcal{F}T \to \mathcal{M}\mathcal{F}T$$

identifies everything to a point.

By duality, then, we get the following information: (For convenience, let G^o stand for $G^o \times \{0\}$ and R^+ for $R^+ \times \{1\}$. C denotes the complex numbers.)

$$\mathcal{P}(T) = \mathcal{C}(T);$$

$$\mathcal{U}(T)|_{G^o} = C, \quad \mathcal{U}(T)|_{R^+} = \mathcal{U}(R^+);$$

$$V(T)|_{G^o} = C, \quad V(T)|_{R^+} = W(R^+);$$

$$W(FT)|_{G^o} = C(G^o) = C_o(R^+) \oplus C,$$

$$W(FT)|_{R^+} = C_o(R^+) \oplus C;$$

$$A(FT)|_{G^o} = C, \quad A(FT)|_{R^+} = C;$$

$$M(FT) = C$$

38. **Remarks**: Restricting the above algebras to the subsemigroup S of T (which is identical to the subsemigroup FS of FT), we obtain information about $LMC(S)$, $LUC(S)$, etc., because of Corollary 34. In particular, we may conclude that all of the containments in the following chain are proper:

$$SAP(S) \subset AP(S) \subset LWP(S) \begin{matrix} \subset WAP(S) \subset \\ \subset LUC(S) \subset \end{matrix} LMC(S)$$

Appendix B

Synopsis

In this appendix we summarize in tabular form many of the important ideas developed in the monograph.

In each table the leftmost column indicates the subspace of $C(S)$ under consideration. In column 2 we list the defining conditions for a function to belong to the given subspace. In many cases alternate conditions are possible; these are discussed in the text. Most of the (canonical) compactifications (ψ,X) listed in column 3 are F-compactifications (Definition II.2.1), which means among other things, that the subspace F of $C(S)$ is in fact a C*-subalgebra, that X is its spectrum, and that the embedding $\psi : S \to X$ is the canonical evaluation mapping. Three of the compactifications in column 4 are, however, F-affine compactifications (Definition II.1.1); namely (cp, cPS), (ck, cKS), and (cq, cQS). For these X is the set of means on F, and ψ is the evaluation mapping. Note than under these circumstances, it is the convex hull $co(\psi(S))$ that is dense in X. With regard to column 4 recall that a compactification (ψ,X) is maximal with respect to a set of properties P if and only if (ψ,X) possesses properties P, and whenever another compactification (ψ_1, X_1) possesses properties P, then there is a continuous homomorphism

$$\phi : X \to X_1$$

such that the diagram

$$\begin{array}{ccc} & S & \\ {\scriptstyle\psi}\swarrow & & \searrow{\scriptstyle\psi_1} \\ X & \rightarrow & X_1 \end{array}$$

commutes.

The fixed point properties in column 5 are the best known and perhaps most important; the list is not exhaustive.

For the category theory in columns 6, 7 and 8, we use the concept of an S-module; that is, a transformation semigroup (S,X) in which the phase space X is itself a semigroup and for which an appropriate distribution law holds. The (phase space of the) S-module may be topological, semitopological, or right topological. Moreover, we consider actions $(s,x) \to s \cdot x : S \times X \to X$ of three types: (1) (jointly) continuous actions, (2) separately continuous actions, and (3) actions which are separately and k-continuous (meaning, continuous on compact subsets of $S \times X$).

In the comments column we have briefly indicated some distinguished property or problem about the space or compactification. For the most classical cases (*SAP*, *AP*, *WAP*) it is impossible in so short a space to begin to indicate their significances and attractions. Even for some of the less well-known situations, much that is in this monograph is not included, even by inference, in this synopsis.

Let S be a semitopological semigroup

(1) Subspace F of C(S)	(2) Condition for a function f to be in F	(3) Compactification of S (III.14.2)
LMC	$s \to \mu(L_s f)$ is continuous for every $\mu \in \beta S$ (III.4.1)	(p,PS)
WLUC	$s \to L_s f$ is weakly continuous (III.3.1)	(cp,cPS)
LUC	$s \to L_s f$ is norm continuous (III.5.1)	(u,US)[1]
K	$R_S f$ is relatively compact-open compact (III.6.1)	(k,KS)
CK	co $R_S f$ is relatively compact-open compact (III.7.1)	(ck,cKS)
WAP	$R_S f$ is relatively weakly compact (III.8.1)	(w,WS)[1]
LWP	$s \to L_s f$ is norm continuous and $R_S f$ is relatively weakly compact (III.11.1)	(v,VS)[1]
KWP	$R_S f$ is relatively compact-open compact and relatively weakly compact (III.12.1)	(q,QS)
CKWP	co $R_S f$ is relatively compact-open compact and relatively weakly compact (III.13.1)	(cq,cQS)
AP	$R_S f$ is relatively norm compact (III.9.1)	(a,AS)[1]
SAP	uniform limit of linear combinations of coefficients of finite dimensional weakly continuous unitary representations (III.10.2)	(m,MS)

[1]An affine compactification also exists.

Let S be a semitopological semigroup.

(1)	(3)	(4)
		Compactification is maximal with respect to the following properties:
LMC	(p,PS)	compact right topological semigroup PS with p(S) dense in PS and $P(S) \subset \Lambda(P(S))$[2]. (III.4.5)
WLUC	(cp,cPS)	compact affine right topological semigroup cPS with the convex hull co(cp(S)) of cp(S) dense in cPS and $cp(S) \subset \Lambda(cPS)$. (III.3.5)
LUC	(u,US)	compact right topological semigroup US with u(S) dense in US, $u(S) \subset \Lambda(US)$, and the function $(s,x) \to u(s)x$ continuous on $S \times US$. (III.5.5)
K	(k,KS)	compact right topological semigroup KS with k(S) dense in KS, $k(S) \subset \Lambda(KS)$, and $(s,x) \to k(s)x$ continuous on $C \times KS$ for every compact $C \subset S$. (III.6.3)
CK	(ck,cKS)	compact affine right topological semigroup cKS with co(k(S)) dense in cKS, $cK(S) \subset \Lambda(cKS)$, and $(s,x) \to ck(s)x$ continuous on $C \times cKS$ for every compact $C \subset S$. (III.7.3)
WAP	(w,WS)	compact semitopological semigroup WS with w(S) dense in WS. (III.8.4)
LWP	(v,VS)	compact semitopological semigroup VS with v(S) dense in VS and the function $(s,x) \to v(s)x$ continuous on $S \times VS$. (III.11.3)
KWP	(q,QS)	compact semitopological semigroup QS with q(S) dense in QS and the function $(s,x) \to q(s)x$ continuous on $C \times QS$ for every compact $C \subset S$. (III.12.3)
CKWP	(cq,cQS)	compact affine semitopological semigroup cQS with co(cq(S)) dense in cQS and the function $(s,x) \to cq(s)x$ continuous on $C \times cQS$ for every compact $C \subset S$. (III.13.3)
AP	(a,AS)	compact topological semigroup AS with a(S) dense in AS. (III.9.4)
SAP	(m,MS)	compact topological group MS with m(S) dense in MS. (III.10.4)

[2] For a right topological semigroup T, $\Lambda(T) = \{y \in T \mid x \to yx$ is continuous$\}$.

Let S be a semitopological semigroup

(1)	(3)	(5) Fixed Point Properties and Left Amenability
LMC	(p,PS)	extremely left amenable ⇔ every separately continuous flow has a fixed point. (IV.2.7)
WLUC	(cp,cPS)	left amenable ⇔ every separately continuous affine flow has a fixed point. (IV.1.9)
LUC	(u,US)	extremely left amenable ⇔ every jointly continuous flow has a fixed point. (IV.2.7;see also IV.1.9)
K	(k,KS)	extremely left amenable ⇔ every separately continuous flow which is equicontinuous on compacta has a fixed point. (IV.2.7)
CK	(ck,cKS)	left amenable ⇔ every separately continuous affine flow which is equicontinuous on compacta has a fixed point. (IV.1.9)
WAP	(w,WS)	extremely left amenable ⇔ every separately continuous quasi-equicontinuous flow has a fixed point. (IV.2.7;see also IV.1.9)
LWP	(v,VS)	extremely left amenable ⇔ every jointly continuous quasi-equicontinuous flow has a fixed point (IV.2.7)
KWP	(q,QS)	extremely left amenable ⇔ every separately continuous quasi-equicontinuous flow which is equicontinuous on compacta has a fixed point. (IV.2.7)
CKWP	(cq,cQS)	left amenable ⇔ every separately continuous quasi-equicontinuous affine flow which is equicontinuous on compacta has a fixed point. (IV.1.9)
AP	(a,AS)	extremely left amenable ⇔ every separately continuous equicontinuous flow has a fixed point (IV.2.7; see also IV.1.9)
SAP	(m,MS)	left amenable always and every separately continuous equicontinuous affine distal flow always has a fixed point.

Let S be a semitopological semigroup with identity.

(1)	(6) Category	(7) Universal map	(8) Universal Properties
LMC	SEP RT S-MOD[3] (A.9)	(p_S, PS)	compact right topological semigroup on which S acts on the left with a separately continuous action.
WLUC	A SEP RT S-MOD (A.11)	(cp_S, cPS)	compact affine right topological semigroup on which S acts on the left with a separately continuous action.
LUC	RT S-MOD	(u_S, US)	compact right topological semigroup on which S acts on the left with a jointly continuous action.
K	RT S-*k*MOD	(k_S, KS)	compact right topological semigroup on which S acts on the left with an action which is separately and *k*-continuous
CK	ART S-*k*MOD	(ck_S, cKS)	compact affine right topological semigroup on which S acts on the left with an action which is separately and *k*-continuous.
WAP	ST SGP	(w_S, WS)[4]	compact semitopological semigroup
LWP	ST S-MOD	(v_S, VS)	compact semitopological semigroup on which S acts on the left with a jointly continuous action.
KWP	ST S-*k*MOD	(q_S, QS)	compact semitopological semigroup on which S acts on the left with an action which is separately and *k*-continuous.
CKWP	AST S-*k*MOD	(cq_S, cQS)	compact affine semitopological semigroup on which S acts on the left with an action which is separately and *k*-continuous.
AP	TOP SGP	(a_S, AS)[4]	compact topological semigroup
SAP	TOP GRP	(m_S, MS)	compact topological group

[3] We require that an S-module T have an identity 1 and that the closure of $S \cdot 1$ in T be a retract of T.

[4] Is also considered as an S-module.

Let S be a semitopological semigroup

(1)	(9) Comments
LMC	*LMC* is the largest left m-introverted subalgebra of C(S). (III.4.4)
WLUC	*WLUC* is the largest left introverted subspace of C(S)(III.3.4). No known example where *WLUC* ≠ *LMC*, (see III.14.9,14.10).
LUC	On a topological group coincides with uniformly continuous functions. (III.5.2) For a strongly countably complete, regular, semitopological group, *LUC* = *LMC*. (III.14.6)
K	On a *k*-space coincides with *LUC* (III.14.8). No known example where *K* ≠ *LUC*.
CK	No known example where *CK* ≠ *K*.
WAP	The original generalization of *AP*. Structure of K(WS) relates naturally to decomposition of *WAP*. Of importance in harmonic analysis.
LWP	A classical result of Eberlein is that for a locally compact group *LWP* = *WAP*.
KWP	For *k*-spaces coincides with *LWP*. No known example where *KWP* ≠ *LWP*.
CKWP	No known example where *CKWP* ≠ *KWP*.
AP	"The literature on almost periodic functions is enormous" - E. Hewitt and K.A. Ross, *Abstract Harmonic Analysis I*, 1963. The space motivating much of the subsequent theory of functions on semigroups.
SAP	For groups coincides with *AP*. On semigroups studied extensively by Maak.

NOTATION

INDEX

REFERENCES

E. M. Alfsen (1971), Compact Convex Sets and Boundary Integrals, Springer, New York.

L. Argabright (1968), Invariant means and fixed points; a sequel to Mitchell's paper, Trans. Amer. Math. Soc. 130, 127-130.

J. W. Baker and R. J. Butcher (1976), The Stone-Čech compactification of a topological semigroup, Proc. Camb. Phil. Soc. 80, 103-107.

J. W. Baker and P. Milnes (1977), The ideal structure of the Stone-Čech compactification of a group, to appear in Proc. Camb. Phil. Soc.

J. F. Berglund (1970), On extending almost periodic functions, Pacific J. Math. 33, 281-289.

J. F. Berglund and K. H. Hofmann (1967), Compact Semitopological Semigroups and Weakly Almost Periodic Functions, Springer, New York.

J. F. Berglund and P. Milnes (1976), Algebras of functions on semitopological left-groups, Trans. Amer. Math. Soc. 222, 157-178.

J. R. Blum and D. L. Hanson (1960), On invariant probability measures I, Pacific J. Math. 10, 1125-1129.

R. B. Burckel (1970), Weakly Almost Periodic Functions on Semigroups, Gordon and Breach, New York, and mimeographed addendum.

R. J. Butcher (1975), The Stone-Čech compactification of a topological semigroup and its algebra of measures, thesis, Sheffield.

C. Chou (1969), Minimal sets and ergodic measures for $\beta N \backslash N$, Illinois J. Math. 13, 777-788.

C. Chou (1970), On topologically invariant means on a locally compact group, Trans. Amer. Math. Soc. 151, 443-456.

C. Chou (1971), On a geometric property of the set of invariant means on a group, Proc. Amer. Math. Soc. 30, 296-302.

C. Chou (1975), Weakly almost periodic functions and almost convergent functions on a group, Trans. Amer. Math. Soc. 206, 175-200.

H. Cohen and H. S. Collins (1959), Affine semigroups, Trans. Amer. Math. Soc. 93, 97-113.

H. S. Collins (1962), Remarks on affine semigroups, Pacific J. Math. 12, 449-455.

W. W. Comfort and K. A. Ross (1966), Pseudocompactness and uniform continuity in topological groups, Pacific J. Math. 16, 483-496.

M. M. Day (1957), Amenable semigroups, Illinois J. Math. 1, 509-544.

M. M. Day (1961), Fixed point theorems for compact convex sets, Illinois J. Math. 5, 585-589, and correction 8(1964), 713.

M. M. Day (1969), Semigroups and amenability, in Semigroups (editor: K. W. Folley), Academic Press.

K. deLeeuw and I. Glicksberg (1961), Applications of almost periodic compactifications, Acta Math. 105, 63-97.

K. deLeeuw and I. Glicksberg (1965), The decomposition of certain group representations, J. Analyse Math. 15, 135-192.

J. Dixmier (1964), Les C*-algèbres et leurs représentations, Gauthier-Villars, Paris.

J. Dugundji (1966), Topology, Allyn and Bacon, Boston.

N. Dunford and J. T. Schwartz (1964), Linear Operators I (second printing), Wiley, New York.

R. Ellis (1969), Lectures on Topological Dynamics, Benjamin, New York.

L. Fairchild (1972), Extreme invariant means without minimal support, Trans. Amer. Math. Soc. 172, 83-93.

H. Furstenberg (1961), Strict ergodicity and transformation of the torus, Amer. J. Math. 83, 573-601.

I. Glicksberg (1959), Convolution semigroups of measures, Pacific J. Math. 9, 51-67.

I. Glicksberg (1961), Weak compactness and separate continuity, Pacific J. Math. 11, 205-214.

E. Granirer (1965 and 1967), Extremely amenable semigroups I and II, Math. Scand. 17, 177-197 and 20, 93-113.

E. Granirer and A. T. Lau (1971), Invariant means on locally compact groups, Illinois J. Math. 15, 249-257.

F. P. Greenleaf (1969), Invariant Means on Topological Groups and Their Applications, Van Nostrand, New York.

F. P. Greenleaf (1973), Ergodic theorems and the construction of summing sequences in amenable locally compact groups, Comm. Pure Appl. Math. 26, 29-46.

A. Grothendieck (1952), Critères de compacité dans les espaces fonctionnels généraux, Amer. J. Math. 74, 168-186.

G. H. Hardy and J. E. Littlewood (1914), Some problems of diophantine approximation, Acta Math. 37, 155-191 (or Collected Papers of G. H. Hardy, volume I, Oxford University Press, 1966).

H. Herrlich and G. E. Strecker (1973), Category Theory: An Introduction, Allyn and Bacon, Boston.

E. Hewitt and K. A. Ross (1963 and 1970), Abstract Harmonic Analysis I and II, Springer, New York.

J. A. Hildebrant and J. D. Lawson (1973). The Bohr compactification of a dense ideal in a topological semigroup, Semigroup Forum 6, 86-92.

K. H. Hofmann and P. S. Mostert (1966), Elements of Compact Semigroups, Merrill, Columbus.

K. Jacobs (1956), Ergodentheorie und fastperiodische Functionen auf Halbgruppen, Math. Z. 64, 298-338.

H. D. Junghenn (1975), Some general results on fixed points and invariant means, Semigroup Forum 11, 153-164.

M. Katětov (1951), On real-valued functions in topological spaces, Fund. Math. 38, 85-91, and correction 40(1953), 203-205.

J. L. Kelley (1955), General Topology, Van Nostrand, New York.

A. T. Lau (1973), Invariant means on almost periodic functions and fixed point properties, Rocky Mountain J. Math. 3, 69-76.

A. T. Lau (1976), Compactifications of semigroups and transformations, preprint.

N. Macri (1974), The continuity of Arens' product on the Stone-Čech compactification of semigroups, Trans. Amer. Math. Soc. 191, 185-193.

P. Milnes (1973), Compactifications of semitopological semigroups, J. Australian Math. Soc. 15, 488-503.

P. Milnes (1975), On the extension of continuous and almost periodic functions, Pacific J. Math. 56, 187-193.

P. Milnes (1976), An extension theorem for functions on semigroups, Proc. Amer. Math. Soc. 55, 152-154.

P. Milnes (1977), Left mean-ergodicity, fixed points and invariant means, to appear in J. Math. Anal. Appl.

P. Milnes and J. S. Pym (1977), Counterexample in the theory of continuous functions on topological groups, Pacific J. Math. 66, 205-209.

T. Mitchell (1965), Constant functions and left invariant means on semigroups, Trans. Amer. Math. Soc. 119, 244-261.

T. Mitchell (1966), Fixed points and multiplicative left invariant means, Trans. Amer. Math. Soc. 122, 195-202.

T. Mitchell (1968), Function algebras, means and fixed points, Trans. Amer. Math. Soc. 130, 117-126.

T. Mitchell (1970), Topological semigroups and fixed points, Illinois J. Math. 14, 630-641.

A. Mukherjea and N. A. Tserpes (1973), Invariant measures and the converse of Haar's theorem on semitopological semigroups, Pacific J. Math. 44, 251-262.

I. Namioka (1972), Right topological groups, distal flows, and a fixed point theorem, Math. Systems Theory 6, 193-209.

I. Namioka (1974), Separate continuity and joint continuity, Pacific J. Math. 51, 515-531.

R. R. Phelps (1966), Lectures on Choquet's Theorem, Van Nostrand, New York.

J. S. Pym (1964), The convolution of linear functionals, Proc. London Math. Soc. 14, 431-444.

J. S. Pym (1965), The convolution of functionals on spaces of bounded functions, Proc. London Math. Soc. 15, 84-104.

J. S. Pym (1968), Idempotent probability measures on compact semitopological semigroups, Proc. Amer. Math. Soc. 21, 499-501.

J. S. Pym (1969), Convolution and the second dual of a Banach algebra, Proc. Camb. Phil. Soc. 67, 597-599.

R. A. Raimi (1964), Minimal sets and ergodic measures in $\beta N \backslash N$, Bull. Amer. Math. Soc. 70, 711-712.

C. R. Rao (1965), Invariant means on spaces of continuous or measurable functions, Trans. Amer. Math. Soc. 114, 187-196.

J. M. Rosenblatt (1976), Invariant means and invariant ideals in $L_\infty(G)$ for a locally compact group G, J. Functional Analysis 21, 31-51.

W. Ruppert (1973), Rechtstopologische Halbgruppen, J. reine angew. Math. 261, 123-133.

W. Ruppert (1974), Rechtstopologische Intervallhalbgruppen und Kreishalbgruppen, Manuscripta Math. 14, 183-193.

W. Ruppert (1975), Über kompakte rechtstopologische Gruppen mit gleichgradig stetigen Linkstranslationen, Anz. Österreich. Akad. Wiss. Math.-naturwiss. Kl. 184, 159-169.

J. von Neumann (1929), Zur allgemeinen Theorie des Masses, Fund. Math. 13, 73-116.

A. Weil (1937), Sur les espaces à structure uniforme et sur la topologie générale, Hermann, Paris.

C. Wilde and K. Witz (1967), Invariant means and the Stone-Čech compactification, Pacific J. Math. 21, 577-586.

K. Witz (1964), Applications of a compactification for bounded operator semigroups, Illinois J. Math. 8, 685-696.